Patent Made Simple: How to File Without a Lawyer

Thomas Heinz Meitinger

Patent Made Simple: How to File Without a Lawyer

Guide to Creating Valuable Patents and Utility Models

Springer

Thomas Heinz Meitinger
Meitinger Patentanwalts GmbH
Munich, Germany

ISBN 978-3-662-73095-9 ISBN 978-3-662-73096-6 (eBook)
https://doi.org/10.1007/978-3-662-73096-6

Translation from the German language edition: "Ohne Anwalt zum Patent" by Thomas Heinz Meitinger, © Der/die Herausgeber bzw. der/die Autor(en), exklusiv lizenziert durch Springer-Verlag GmbH, DE, ein Teil von Springer Nature 2024. Published by Springer Berlin Heidelberg. All Rights Reserved.

This book is a translation of the original German edition "Ohne Anwalt zum Patent" by Thomas Heinz Meitinger, published by Springer-Verlag GmbH, DE in 2021. The translation was done with the help of an artificial intelligence machine translation tool. A subsequent human revision was done primarily in terms of content, so that the book will read stylistically differently from a conventional translation. Springer Nature works continuously to further the development of tools for the production of books and on the related technologies to support the authors.

This Springer imprint is published by the registered company Springer-Verlag GmbH, DE, part of Springer Nature.
The registered company address is: Heidelberger Platz 3, 14197 Berlin, Germany

Preface

In my many years of practice as a patent attorney, I have experienced several times that good ideas did not lead to good patents. The reason was often that inventors found the cost of a patent attorney too high. In financial distress, patent or utility model applications were written by the inventors themselves and submitted to the patent office.

If these patent specifications written by the inventor themselves were infringed by a third party, they often could not be used to stop the imitator due to their poor quality. License fees could also not be obtained. These "do-it-yourself" applications simply were not worth the effort.

This book is a guide for inventors on how to write patent and utility model applications themselves. With the knowledge provided in this book, well-formulated patents and utility models can be obtained, enabling the inventor to secure effective protection for their intellectual property. Nothing will then stand in the way of the inventor's successful commercial exploitation of their invention.

Munich
in May 2021

Patent Attorney Dr. Thomas Heinz Meitinger

Laws

Employee Inventions Act: Act on Employee Inventions in the consolidated version published in the Federal Law Gazette Part III, classification number 422-–1, as last amended by Article 7 of the Act of July 31, 2009 (Federal Law Gazette I p. 2521).

BGB: German Civil Code in the version promulgated on January 2, 2002 (Federal Law Gazette I p. 42, 2909; 2003 I p. 738), last amended by Article 10 of the Act of March 30, 2021 (Federal Law Gazette I p. 607).

EPC: European Patent Convention as amended by the Act revising the Convention of 29 November 2000 and the decision of the Administrative Council of 28 June 2001 adopting the revised text of the European Patent Convention (OJ EPO, Special Editions No. 4/2001, pages 56 ff.; No. 1/2003, pages 3 ff.; No. 1/2007, pages 1–88) and the Implementing Regulations as amended by the decision of the Administrative Council of 7 December 2006 (OJ EPO, Special Edition No. 1/2007, pages 89 ff.), as subsequently amended by the decisions of the Administrative Council of 6 March 2008 (OJ EPO 2008, 124), 21 October 2008 (OJ EPO 2008, 513), 25 March 2009 (OJ EPO 2009, 296 and 299), 27 October 2009 (OJ EPO 2009, 582), 28 October 2009 (OJ EPO 2009, 585), 26 October 2010 (OJ EPO 2010, 568, 634 and 637), 27 June 2012 (OJ EPO 2012, 442), 16 October 2013 (OJ EPO 2013, 501 and 503), 13 December 2013 (OJ EPO 2014, A3 and A4), 15 October 2014 (OJ EPO 2015, A17), 14 October 2015 (OJ EPO 2015, A82 and A83), 30 June 2016 (OJ EPO 2016, A100), 14 December 2016 (OJ EPO 2016, A102), 28 June 2017 (OJ EPO 2017, A55), 29 June 2017 (OJ EPO 2017, A56), 13 December 2017 (OJ EPO 2018, A2), 28 June 2018 (OJ EPO 2018, A57), 28 March 2019 (OJ EPO 2019, A31), 12 December 2019 (OJ EPO 2020, A5) and 27 March 2020 (OJ EPO 2020, A36).

Utility Model Act as promulgated on August 28, 1986 (Federal Law Gazette I p. 1455), last amended by Article 10 of the Act of July 17, 2017 (Federal Law Gazette I p. 2541).

Utility Model Ordinance of May 11, 2004 (Federal Law Gazette I p. 890), last amended by Article 3 of the Ordinance of December 12, 2018 (Federal Law Gazette I p. 2446).

GVG: Courts Constitution Act as promulgated on May 9, 1975 (Federal Law Gazette I p. 1077), last amended by Article 4 of the Act of March 9, 2021 (Federal Law Gazette I p. 327).

Patent Act as promulgated on December 16, 1980 (Federal Law Gazette 1981 I p. 1), last amended by Article 4 of the Act of October 8, 2017 (Federal Law Gazette I p. 3546).

Patent Costs Act of December 13, 2001 (Federal Law Gazette I p. 3656), last amended by Article 3 of the Act of December 11, 2018 (Federal Law Gazette I p. 2357).

Patent Ordinance of September 1, 2003 (Federal Law Gazette I p. 1702), last amended by Article 1 of the Ordinance of December 12, 2018 (Federal Law Gazette I p. 2446).

PCT: Patent Cooperation Treaty signed in Washington on June 19, 1970, amended on September 28, 1979, February 3, 1984, and October 3, 2001.

Paris Convention/CUP/PVÜ: Paris Convention for the Protection of Industrial Property of March 20, 1883, revised in BRUSSELS on December 14, 1900, in WASHINGTON on June 2, 1911, in THE HAGUE on November 6, 1925, in LONDON on June 2, 1934, in LISBON on October 31, 1958, and in STOCKHOLM on July 14, 1967, and amended on October 2, 1979.

ZPO: Code of Civil Procedure as amended by the announcement of December 5, 2005 (Federal Law Gazette I p. 3202; 2006 I p. 431; 2007 I p. 1781), last amended by Article 8 of the Act of December 22, 2020 (Federal Law Gazette I p. 3320).

Contents

About the Author

Dr. Thomas Heinz Meitinger is a German and European patent attorney. He is the managing director of Meitinger Patentanwalts GmbH. Meitinger Patentanwalts GmbH is a medium-sized patent law firm based in Munich. After studying electrical engineering in Karlsruhe, he initially worked as a development engineer. Subsequent positions included roles as production manager and technical director in medium-sized companies. Dr. Meitinger regularly publishes scientific articles, writes specialist books on intellectual property law, and gives lectures on topics related to patent and trademark law. Dr. Meitinger holds the degrees Dipl.-Ing. (Univ.) and Dipl.-Wirtsch.-Ing. (FH). In addition, he holds the following master's degrees: LL.M., LL.M., MBA, MBA, M.A., and M.Sc.

Abbreviations

ABl. EPA	Official Journal of the European Patent Office
BGH	Federal Court of Justice (Bundesgerichtshof)
BPatG	Federal Patent Court (Bundespatentgericht)
DPMA	German Patent and Trademark Office (Deutsches Patent- und Markenamt)
EPO/EPA	European Patent Office (Europäisches Patentamt)
EPC/EPÜ	European Patent Convention (Europäisches Patentübereinkommen)
PCT	Patent Cooperation Treaty
WIPO	World Intellectual Property Organization
ZPO	Code of Civil Procedure (Zivilprozessordnung)

List of Figures

List of Tables

1 Introduction

Suddenly it's there: the brilliant idea. Naturally, you want to help it succeed. To do this, the idea must be promoted and presented in many presentations until the breakthrough finally comes.

However, there is a significant risk that commercial exploitation will occur without involving the inventor in the success. A patent or a utility model prevents abusive exploitation. With an intellectual property right, the inventor can prevent any third party from exploiting their invention without permission. Therefore, an inventor should apply for a patent or a utility model for their invention.

According to the Patent Act, an inventor is the person who creates an invention. The inventor is entitled to all rights to the invention.[1] In addition, the Patent Act recognizes the applicant. The applicant files the invention for a patent or a utility model. The Patent Office initially assumes that the applicant is entitled to file the invention as a patent for themselves.[2]

The preparation of a utility model or patent application can be commissioned from a patent attorney. However, this may incur costs that deter an inventor. An alternative is to draft a patent application oneself and submit it to the Patent Office.

This technical book is a guide to independently preparing a patent or utility model application. The individual components of an application are described, and suggested formulations are provided for each. Numerous practical examples offer additional guidance.

[1] Section 6, sentence 1 of the Patent Act.

[2] Section 7, paragraph 1 of the Patent Act.

T. H. Meitinger, *Patent Made Simple: How to File Without a Lawyer*,
https://doi.org/10.1007/978-3-662-73096-6_1

The structure of the technical book corresponds to the recommended procedure for preparing an application. First, the state of the art[3] is explained, which represents the starting point for the creation of the invention. Through engagement with the state of the art, the inventor identifies disadvantages that prompt them to make their invention. These disadvantages lead to the objective of the invention. The next step in preparing an application is a description of the various embodiments of the invention. Finally, the claims are drafted.

This procedure has the advantage that the most difficult step—formulating the claims—occurs only at the end of the process. By then, the inventor has a profound understanding of the various aspects of their invention, which makes drafting the claims considerably easier.

Many real-world examples from practice are presented. Improved formulations or representations are provided for these examples. The examples are drawn in particular from the technical fields of bicycles and children's toys. These technical fields generally involve simple technical teachings. Understanding the formulation of a patent or utility model application should not be unnecessarily complicated by complex technical matters.

The most challenging part of preparing a patent or utility model application is drafting the claims. Using the instructions in this manual, it should be possible to produce at least a usable initial set of claims. It is important that the claims are formulated in such a way that they can initiate a targeted examination procedure. A perfect set of claims cannot be expected solely on the basis of reading a book. However, this is not necessary, since the purpose of the grant procedure is precisely to achieve this. Through the examination procedure before the Patent Office, the applicant should arrive at a set of claims that grants them the broadest possible scope of protection.

In Chap. 2 tools are presented to help decide whether it is worthwhile to file a patent or utility model at all. Not every idea stands up to scrutiny for economic significance. In such cases, one should save oneself the effort and cost of a patent or utility model. The third chapter then provides an overview of the various stages in the "life" of a patent or utility model. This is followed in Chap. 4 by general formal requirements that must be observed to avoid an official notice from the Patent Office due to formal deficiencies. In Chap. 5 an overview of the individual components of an application is provided. These individual components are explained in detail in Chapters 6 to 10.

In Chap. 11 the application forms of the Patent Offices are presented, which must be attached to the description of the invention and the claims. Chapter 12 is dedicated to

[3] "The state of the art comprises all knowledge made available to the public before the date relevant for the priority of the application by written or oral description, by use, or in any other way." (Section 3, paragraph 1, sentence 2 of the Patent Act) The priority date of a patent or utility model application is the filing date. If the priority of an earlier application has been claimed, the priority date is the filing date of the earlier application.

the various ways of conducting a search for the state of the art. Chapter 13 serves as guidance for inventors who are in an employment relationship. Subsequently, Chap. 14 discusses the key features of an inventor or applicant group, since in practice inventions are often created in teams. Chaps. 15, 16, 17, and 18 then present the different ways to obtain an intellectual property right. The various options are: a German utility model, a German patent, a European patent, or an international patent application. In Chap. 19 the response to an official notice in an examination or registration procedure is addressed, and the most common deficiency notices are discussed. The final two chapters present and explain templates for a patent or utility model application and a response to an official notice.

When drafting the application documents, it should be kept in mind that the initial recipient of the patent or utility model application is an examiner at the Patent Office. The application should be formulated in such a way that it makes it as easy as possible for the examiner to grant the patent. Accordingly, the invention should be described in detail and clearly. It may be advantageous to use appropriate wording to highlight the patentability of the invention. For example, it is preferable to refer to a "technical" problem of the invention rather than just a problem. The phrase "the present invention" emphasizes that the application concerns a patentable invention. It is advisable to refer to technically advantageous embodiments of the invention in the dependent claims. With appropriate wording, it may be possible to support an examiner's existing inclination to grant a patent. This can at least accelerate the granting of a patent.

The book presents numerous practical examples and discusses them in commentary. The examples are taken from patent and utility model publications, whose publication numbers are provided. The excerpts from the patent and utility model publications are written in italics. Individual passages in these excerpts are written in bold and underlined, and in square brackets behind these passages are [Point 1], [Point 2], [Point 3], etc. Comments follow these points. In addition, the book contains tips that concisely summarize the essential content of the relevant sections of the technical book.

2 Why a Patent or Utility Model?

Patents and utility models are industrial property rights that protect inventive devices and processes from imitation. To this end, a patent or utility model provides an exclusive right to prohibit. Accordingly, the use of any products or processes by an unauthorized third party that fall within the protected scope of the patent can be prohibited by the rights holder. A patent or utility model is an economic monopoly right. However, a patent or utility model only has actual value if the protected subject matter is economically valuable.

One should avoid the misconception of seeing a patent or utility model as inherently valuable. The granting of a patent or the registration of a utility model is not an indication of a valuable invention. The patent office examines novelty and inventive step, but not economic value. However, the theoretical feasibility is examined by the patent office before patenting.[1] After a patent has been granted, it can be assumed that the invention is at least theoretically feasible. For a utility model, only a formal examination for defects is carried out before registration.

The effort of filing a patent or utility model application only makes sense if the invention is actually expected to have economic value.

Tip: You should only undertake the effort of filing a patent or utility model application for an invention that is economically valuable.

Unfortunately, an invention cannot be filed for a patent only after it has proven to be economically valuable. Even a single sale of a product that implements the invention[2]

[1] Section 34, paragraph 4 of the Patent Act.

[2] Decisions of the Boards of Appeal of the EPO: T 482/89, OJ EPO 1992, 646 and T 1022/99 (The EPO designates decisions of its technical Boards of Appeal with a leading "T").

T. H. Meitinger, *Patent Made Simple: How to File Without a Lawyer*,
https://doi.org/10.1007/978-3-662-73096-6_2

means the invention is no longer considered novel and is thus no longer patentable.[3] It is therefore necessary to assess the economic significance of the invention before its actual implementation.

2.1 Market Potential

The economic value of a patent is determined in particular by the expected market potential of the underlying invention. Consideration of one's own capabilities for production or distribution of a product based on the invention can initially be disregarded, since a patent can be licensed out.

When assessing market potential, the following three criteria in particular can be considered: market situation, competitive situation, and customer situation. The market situation can be considered positive if the market has a high market volume. It is also advantageous for the prospects of an invention if the relevant market is characterized by high growth. If the market does not have an absolutely dominant market leader, this can also be advantageous for the economic assessment of an invention. In this case, a newcomernamely, the patent holder can enter the market more successfully. If the customer situation in a market is characterized by high heterogeneity among potential customers, this can also promote the success of an invention in that market.

2.2 Priority Period

The one-year priority period[4] begins immediately after the filing of the application documents.[5] During the priority period, a so-called subsequent application for the same invention can be filed in any country in the world. This subsequent application receives the same early priority date as the first application.[6] However, the same priority date only applies to the same subject matter. If new subject matter is included in the subsequent

[3] This deficiency can be remedied by the novelty grace period under the Utility Model Act. Section 3, paragraph 1, sentence 3 of the Utility Model Act provides for a general novelty grace period: "A disclosure or use that occurred within six months prior to the date relevant for the priority of the application shall not be taken into account if it is based on the applicant's or his predecessor's work." However, this novelty grace period should only be regarded as a last resort.

[4] The basic right of priority for patent or utility model applications is codified in Article 4A, paragraph 1 of the Paris Convention (foreign priority). The one-year period within which the right of priority can be claimed is specified in Article 4C, paragraph 1 of the Paris Convention.

[5] The priority period begins on the day on which the application is accorded a filing date pursuant to Section 35, paragraph 1 of the Patent Act.

[6] Article 4A, paragraph 1 of the Paris Convention.

application, this new subject matter receives the filing date on which the subsequent application was actually filed. The right of priority thus makes it possible to claim an invention through a single, for example German, patent application and to extend this claim to any countries within one year.

There is also the so-called "internal" priority. During the first year after filing the application documents, a subsequent application can be filed not only abroad but also in Germany. If the inventor comes up with further important embodiments, he can expand the previous application and, by filing the expanded application, claim the priority of the earlier application. For those parts of the later application that were already described in the first application, the filing date of the first application applies. For the remaining parts, the later filing date applies. The claiming of an internal priority is possible for a German patent application[7] and for a German utility model[8]. However, the earlier patent application or utility model is then deemed withdrawn. It is also possible to first file a utility model and then a patent application, or vice versa. In this case, the fiction of withdrawal does not apply, since a utility model can exist in parallel with a patent.[9]

A first patent or utility model application can thus create an initial property right, which can potentially be "corrected" to some extent by a subsequent application. The first application could therefore be prepared cost-effectively by the applicant themselves, and the subsequent application could be professionally prepared by a patent attorney. However, the initial application, whether a patent or utility model application, must meet a minimum quality standard. Otherwise, one is lulled into a false sense of security. This book provides guidance to help achieve the required minimum qualitative standard.

2.3 Publication Period

The publication period stipulates that an application will not be published before the expiry of 18 months.[10] The only exception is if a patent is granted before the expiry of the 18 months. For the first one and a half years, an application is therefore kept secret

[7] Section 40, paragraph 1 of the Patent Act.

[8] Section 6, paragraph 1, sentence 1 of the Utility Model Act.

[9] Section 40, paragraph 5, sentence 1 of the Patent Act: by filing a later patent application, the first patent application is deemed withdrawn (fiction of withdrawal) if the later application claims the priority of the first patent application. If the later application is a utility model application, the patent application remains in force (Section 40, paragraph 5, sentence 2 of the Patent Act). There is also an internal priority for the Utility Model Act (Section 6, paragraph 1, sentence 1 of the Utility Model Act). The fiction of withdrawal also applies to two utility model applications (Section 6, paragraph 1, sentence 2 of the Utility Model Act in conjunction with Section 40, paragraph 5, sentence 1 of the Patent Act).

[10] Section 31, paragraph 2, no. 2 of the Patent Act or Article 93, paragraph 1, letter a of the EPC.

by the patent office. This also means that an inventor has 18 months to consider whether the application, as filed, has any prospect of being granted.[11] The inventor can still decide not to pursue a patent, but instead to keep the invention as company know-how. This is particularly relevant if the applicant finds that, although the registered invention is economically interesting, it is unlikely to meet the initial high expectations. In this case, the invention can continue to be used as confidential company know-how without incurring the high costs of further patent prosecution.

If the inventor discovers through a search that his invention is not novel, he can withdraw the application to prevent competitors from learning how his invention works. It can therefore make sense to prepare and file an initial patent or utility model application with the patent office and then continue researching to assess the prospects of a patent being granted.

2.4 Poor Patents Harm, Good Patents Benefit

If the assessment of market potential yields a satisfactory result, one can consider drafting a patent or utility model application. However, it must be borne in mind that an application is only legally valid if the invention is described in such a way that a skilled person can carry it out.[12] From an application, any competitor can therefore learn how the described invention is implemented. This makes effective protection against imitation and circumvention solutions all the more important. Such protection can only be achieved through a well-drafted application and, in particular, professionally prepared claims.

Tip: A well-drafted patent protects. A poorly drafted patent harms.

2.5 Effects

A patent or utility model does not constitute a license to use.[13] The patent office does not check whether there are earlier rights that could prohibit the patent holder from using his patented invention. However, in a patent grant procedure, the patent office does examine whether the present application is novel and inventive over earlier rights.

[11] Section 32, paragraph 4 of the Patent Act: The publication document is also published if the application has been withdrawn, provided that the technical preparations for publication have been completed. If an application is not to be published, the withdrawal should therefore take place at least six to eight weeks before the expiry of the 18-month period.

[12] Section 34, paragraph 4 of the Patent Act or Article 83 of the EPC.

[13] The wording of Section 9, sentence 1 of the Patent Act is misleading: "The patent has the effect that only the patent holder is authorized to use the patented invention within the scope of the applicable law." In fact, a patent is a right to prohibit. Section 9, sentence 2 of the Patent Act therefore clarifies: "Any third party is prohibited …".

2.5.1 Significance of the Filing Date

Filing a patent or utility model application has the effect that anyone who files the same invention after the filing date of the patent or utility model application will come away empty-handed. There is no second place in industrial property protection. The first application for an invention is the only one that stands. For this reason, an early filing date is so crucial.

2.5.2 Effect of Publication

The patent office publishes an application as a so-called "publication document".[14] Publication is made electronically in the patent register.[15] The publication of the patent application takes place after 18 months have elapsed since the filing of the application documents.[16] The application is not published if the patent is granted during the 18-month period.[17] In this case, only the patent specification is published.

Publication gives rise to a right to compensation against anyone who uses the subject matter of the application.[18] The subject matter of the application is initially the scope of protection defined by the claims.[19] However, in the end, the granted claims are decisive for determining entitlement to compensation.[20] The right to compensation can even be retroactively eliminated entirely if the application is rejected by the patent office or withdrawn by the applicant.[21]

2.5.3 Effect of the Granted Patent

A granted patent has the effect that the patent holder can prohibit any third party from, without his consent, manufacturing, offering, placing on the market, using, importing, or possessing a product that is the subject of the patent. The subject matter of the patent is defined by the scope of protection of the patent.[22] If the subject matter of the patent is a

[14] Section 32, paragraph 1, sentence 1, no. 1 of the Patent Act.
[15] Section 32, paragraph 1, sentence 2 of the Patent Act.
[16] Section 31, paragraph 2, no. 2 of the Patent Act.
[17] Section 32, paragraph 2, sentence 2 of the Patent Act.
[18] Section 33, paragraph 1 of the Patent Act.
[19] Section 14, sentence 1 of the Patent Act.
[20] Federal Court of Justice, X ZR 79/04, "Extracoronales Geschiebe", *Gewerblicher Rechtsschutz und Urheberrecht,* 2006, 570.
[21] Section 58, paragraph 2 of the Patent Act.
[22] Section 14, sentence 1 of the Patent Act.

process, the patent holder is entitled to prohibit a third party from using the process. The right to prohibit also extends to a product that is made directly by a patented process and is offered, placed on the market, or imported.[23] The effect of a European patent is equivalent to that of a national patent.[24]

In addition to the right to injunctive relief[25], the patent holder can assert a claim for damages. The prerequisite for this is that the patent infringement is intentional or negligent.[26] Patent infringement is almost always at least negligent, since it is assumed that a patent infringer has failed to exercise the necessary care if he has not informed himself about the existing patent situation. Damages can be calculated in three ways: by analogy to a license, by calculating the infringer's profit, or by the loss of profit. The patent holder is entitled to choose the method of calculation. In the past, the infringer's profit was rarely claimed, as the infringer would regularly understate his profits. This abuse has been at least partially curbed by the Federal Court of Justice, so that it has now become more attractive for the patent holder to claim the infringer's profit.[27] In the license analogy, a license rate is used that corresponds to market conditions. In this case, the infringer must pay the patent holder the turnover multiplied by the license rate. Alternatively, the patent holder can demand compensation for the profit he lost as a result of the patent infringement. In addition to damages, the patent holder can demand the surrender of unjust enrichment.[28]

There is also a right to information, which entitles the patent holder to obtain information about the origin and distribution channels of the infringing products.[29] For this purpose, the patent holder must be given the names and addresses of the manufacturers, suppliers, and previous owners. In addition, information must be provided about the quantity of products manufactured, delivered, received, or still ordered.

Another right of the patent holder is the right to destruction.[30] The patent holder can demand that infringing products in the possession of the infringer be destroyed. Destruction does not take place if the patent infringement can be remedied in another way or if destruction would be disproportionate.

Finally, the patent holder also has a right to recall.[31] With this right, infringing products can be permanently removed from the distribution chain. However, the right to recall only applies to the extent that it is proportionate.

---

[23] Section 9, sentence 2, no. 3 of the Patent Act.

[24] Article 64, paragraph 1 of the EPC.

[25] Section 139, paragraph 1, sentence 1 of the Patent Act.

[26] Section 139, paragraph 2, sentence 1 of the Patent Act.

[27] Federal Court of Justice, I ZR 246/98, "Gemeinkostenanteil", *Gewerblicher Rechtsschutz und Urheberrecht,* 2001, 329.

[28] Section 812, paragraph 1, sentence 1, 2nd alternative of the German Civil Code (BGB).

[29] Section 140b, paragraph 1 of the Patent Act.

[30] Section 140a, paragraph 1, sentence 1 of the Patent Act.

[31] Section 140a, paragraph 3, sentence 1 of the Patent Act.

The patent holder can enforce his rights through legal action before the ordinary courts.[32] Alternatively, he can seek a preliminary injunction. However, a preliminary injunction can usually only be obtained in cases of simple technical matters and clear forms of infringement.

A patent holder should only seek a preliminary injunction in clear cases. Typically, the preliminary injunction is issued without hearing the opposing party. The opposing party can file an objection to the preliminary injunction. In addition, the opposing party must comply with the preliminary injunction, i.e., must refrain from the allegedly infringing acts, which can result in significant economic damage for the opposing party. After the objection is filed, an oral hearing is scheduled at which both parties can present their arguments and facts. If it turns out that the preliminary injunction was unjustified, the patent holder must compensate the opposing party for the entire damage, even if he is not at fault.[33]

Intentional patent infringement constitutes a criminal offense.[34] Intent, as a prerequisite for the legal consequence of this paragraph, is difficult to prove. Moreover, the patent holder generally has no interest in proving a criminal offense by the infringer. The patent holder wants the infringement to cease in the future and to be compensated for his loss. Accordingly, criminal prosecution of patent infringers has so far been of no significance.

2.5.4 Effect of the Utility Model

A registered utility model has the same effects as a patent. This is also the reason why case law requires the same inventive step for a utility model as for a patent.[35] The holder of a utility model is entitled to a right to injunctive relief[36], a claim for damages[37], a right to information,[38] a right to recall[39] and a right to destruction[40] against the third party infringing the utility model.

---

[32] Section 143, paragraph 2, sentence 1 of the Patent Act: For matters of industrial property protection, specialized regional courts or higher regional courts are responsible in each federal state. These are called patent litigation chambers. Patent litigation chambers are located in Dsseldorf, Munich, Mannheim, Berlin, Braunschweig, Erfurt, Frankfurt, Hamburg, Leipzig, Nuremberg, Magdeburg, and Saarbrcken.

[33] Section 945 of the Code of Civil Procedure.

[34] Section 142, paragraph 1 of the Patent Act.

[35] Federal Court of Justice, X ZB 27/05, "Demonstrationsschrank", *Gewerblicher Rechtsschutz und Urheberrecht,* 2006, 842.

[36] Section 24, paragraph 1, sentence 1 of the Utility Model Act.

[37] Section 24, paragraph 2, sentence 1 of the Utility Model Act.

[38] Section 24b, paragraph 1 of the Utility Model Act.

[39] Section 24a, paragraph 2 of the Utility Model Act.

[40] Section 24a, paragraph 1, sentence 1 of the Utility Model Act.

2.5.5 Exceptions to the Effects

The effects of a patent do not extend to acts carried out in the private sphere for non-commercial purposes.[41] In addition, a prior user cannot be prohibited from continuing to use the invention in the future. Prior use exists if, before the application, a third party in Germany had already put the invention into use or at least made the necessary preparations to do so. This third party is entitled to use the invention in his business. The effect of the patent does not apply to the entitled third party.[42]

2.6 Caution: Psychological Effects

It is important to be able to take a critical look at one's own invention. Otherwise, an invention cannot be objectively assessed for its economic value. Not every idea should be immediately filed for a patent. Filing economically insignificant inventions for a patent or utility model is a waste of one's own resources. Therefore, one should maintain a critical distance from one's own invention in order to make a rational business decision as to whether to file a patent application. This necessary critical distance to one's own invention can be disrupted by particular psychological effects. Knowing these psychological mechanisms can help avoid falling prey to them.

2.6.1 Endowment Effect

The discoverer of the endowment effect, Daniel Kahneman, demonstrated this effect by forming two groups, giving each member of the first group a mug. He then asked the members of the first group for what price they would be willing to sell their mug. Kahneman specified a price range between $9.25 and $0.25. The members of the second group, who had not received a mug, were asked for what price they would be willing to buy a mug. The second group was given the same price range of $9.25 to $0.25. Remarkably, the selling price at which the participants in the first group were willing to sell their mug differed significantly from the purchase price of the second group. On average, the selling price was $7.12 and the purchase price was $2.87.[43] Kahneman concluded that a good is valued differently depending on whether one owns it or not. Applied to inventions as intellectual property, it can therefore be assumed that an inventor values his invention more highly than a person who did not create the invention. This endowment

[41] Section 11, no. 1 of the Patent Act.

[42] Section 12, paragraph 1, sentences 1 and 2 of the Patent Act.

[43] Daniel Kahneman, Jack L. Knetsch, and Richard H. Thaler, 'Experimental Tests of the Endowment Effect and the Coase Theorem', *Journal of Political Economy,* 1990, 98, 6, 13251348.

effect may also influence an inventor and lead him to an incorrect, namely too high, valuation of his invention.

2.6.2 Loss Aversion

The loss aversion effect leads to a reluctance to give up or lose something. Discontinuing the pursuit of an invention can be such a surrender. An inventor should therefore be aware that giving up an invention may require more effort than continuing to pursue it.[44] Nevertheless, holding on to economically worthless inventions is a mistake, as it wastes resources that are then no longer available for valuable inventions. An inventor should therefore be aware of the loss aversion effect.

2.6.3 Not-Invented-Here Syndrome

The not-invented-here syndrome states that companies tend to value their own inventions more highly than those of others. In particular, it has been found that companies tend to prefer their own inventions and dismiss those of others as less significant.[45] The not-invented-here effect shows that it is difficult to be critical of one's own inventions. This limited objectivity carries the risk of filing patent applications for inventions that do not deserve it. One should therefore be aware that the risk of wasting resources on uneconomical inventions is always present.

2.7 Success Stories

Many utility models and patents have formed the basis for "success stories." Two will be described as examples.

2.7.1 Bionade the Brewed Lemonade

Dieter Leipold was a certified brewmaster and began his career at Würzburger Bürgerbräu in lemonade production. From the very beginning, he was dissatisfied with the fact that many chemical additives and large amounts of sugar were used in lemonade production.

[44] Daniel Kahneman, Jack L. Knetsch, and Richard H. Thaler, 'Anomalies: The Endowment Effect, Loss Aversion, and Status Quo Bias', *Journal of Economic Perspectiv,* 1991, 5, 1, 193206.

[45] Katrin Hussinger and Annelies Wastyn, 'In Search for the Not-Invented-Here Syndrome: The Role of Knowledge Sources and Firm Success', *R&D Management,* 2015.

Leipold set himself the technical challenge of creating a lemonade that would comply with the purity law of beer brewing. After many years of experimentation, Leipold was able to file a patent for his process on April 12, 1990 (see Fig. 2.1, DPMA, https://depatisnet.dpma.de/DepatisNet/depatisnet?action=pdf&docid=DE000004012000A1&xxxfull=1, accessed on 23 April 2026).

(19) BUNDESREPUBLIK DEUTSCHLAND

DEUTSCHES PATENTAMT

(12) **Offenlegungsschrift**

(11) **DE 4012000 A1**

(51) Int. Cl.⁵:
A 23 L 2/38
A 23 F 3/16
A 23 F 3/34
A 23 F 5/24
C 12 S 3/10
C 12 G 3/00

(21) Aktenzeichen: P 40 12 000.7
(22) Anmeldetag: 12. 4. 90
(43) Offenlegungstag: 18. 10. 90

DE 4012000 A1

(30) Innere Priorität: (32) (33) (31)
12.04.89 DE 39 12 041.4

(71) Anmelder:
Privatbrauerei Peter, 8745 Ostheim, DE

(74) Vertreter:
Jaeger, K., Dipl.-Chem. Dr.rer.nat.; Köster, H., Dipl.-Chem. Dr.rer.nat., Pat.-Anwälte, 8035 Gauting

(72) Erfinder:
Leipold, Dieter, Dipl.-Braumeister, 8745 Ostheim, DE

(54) Alkoholfreies Erfrischungsgetränk

Es wird ein alkoholfreies Erfrischungsgetränk auf Basis von reinen Naturprodukten bereitgestellt. Dieses Erfrischungsgetränk ist beispielsweise dadurch erhältlich, daß man einen per se bekannten Malztrunk oder einen wäßrigen, eine Stickstoffquelle für die Ernährung von Mikroorganisment enthaltenden Extrakt von toxikologisch und lebensmittelchemisch unbedenklichen Pflanzenteilen von einer Pflanze oder von mehreren Pflanzen mit Glucose oder einem glucosehaltigen Zucker versetzt,
die so erhaltene zuckerhaltige Lösung mit einem Mikroorganismus versetzt, der die Fähigkeit besitzt, Glucose in Gluconsäure umzuwandeln,
unter zumindest gelegentlichem Belüften fermentiert,
die Fermentation stoppt
und gewünschtenfalls die fermentierte Lösung filtriert und/oder karbonisiert.
Das erfindungsgemäße Erfrischungsgetränk kann auch aus einem koffeinhaltigen pflanzlichen Kwass bestehen, der gegebenenfalls mit zu 20 Vol.-% Wasser verdünnt sein kann, karbonisiert ist und vor dem Verdünnen einen pH-Wert von kleiner als 3,0 aufweist.

DE 4012000 A1

BUNDESDRUCKEREI 08. 90 008 042/572 17/60

Fig. 2.1 Bionade the brewed lemonade (DE 4012000 A1)

However, it was not until 1998 that the economic breakthrough occurred. A major beverage wholesaler in Hamburg was won over for the product. By 2002, around 200 million bottles were already being produced.[46] Dieter Leipold thus devoted almost his entire career to his invention, demonstrating great perseverance. In the end, the effort and patience paid off.

2.7.2 Copying by Xerography

Chester Carlson (born on 8 February 1906 in Seattle, Washington and died on 19 September 1968 in New York) was the only child of his parents. Due to his father's chronic illnesses, the family lived in abject poverty. Even as a child, Chester Carlson had to work to contribute to the family's livelihood. Eventually, he earned the lion's share of the living expenses for the three-person family. In addition, he managed to finance a physics degree at the California Institute of Technology (CalTech). After graduation, Carlson took a job in the patent department of a company. He also studied part-time at New York Law School, where he earned a diploma as a patent attorney.

While working in the patent department, Carlson noticed the technical problem that copies of printed documents could only be produced laboriously and time-consumingly using a wet process. Chester Carlson set himself the technical challenge of producing copies of documents quickly and cost-effectively. After years of research, Carlson found a solution, which he called electrophotography (electron photography), and filed a patent for his invention on September 8, 1938. Many further patent applications followed, covering improvements to the original idea (see Figs. 2.2 and 2.3, DPMA, https://depatisnet.dpma.de/DepatisNet/depatisnet?action=pdf&docid=US000002221776A&xxxfull=1, accessed on 23 April 2026). Since Chester Carlson was himself a patent attorney, he was able to draft "good" patent applications on his own. The commercial breakthrough did not come until 1960 with the Xerox 914 copier. This laid the foundation for the global corporation Xerox, which employed about 130,000 people in 2012. Carlson died in 1968 at the age of 62, leaving a fortune of around 150 million US dollars.[47] Chester Carlson proved that an invention and well-drafted patents could change the world.

[46] Bettina Weiguny, "Bionade. Eine Limo verndert die Welt" [Bionade. A Lemonade Changes the World], 2009.

[47] George A. W. Boehm and A. Groner, "Science in the Service of Mankind: the Battelle Story", 1972.

Patented Nov. 19, 1940 **2,221,776**

UNITED STATES PATENT OFFICE

2,221,776

ELECTRON PHOTOGRAPHY

Chester F. Carlson, Jackson Heights, N. Y.

Application September 8, 1938, Serial No. 228,905

11 Claims. (Cl. 95—5)

This invention relates to photography.

An object of the invention is to improve photographic processes and devices.

Other objects of the invention will be apparent from the following description and accompanying drawings taken in connection with the appended claims.

The invention comprises the features of construction, combination of elements, arrangement of parts, and methods of manufacture and operation referred to above or which will be brought out and exemplified in the disclosure hereinafter set forth, including the illustrations in the drawings.

In the drawings:

Figure 1 is a sectional elevation of an electrophotographic device;

Figure 2 is a section on the line 2—2 of Figure 1;

Figure 3 shows a modified part;

Figure 4 shows part of a modified device suitable for contact printing;

Figure 5 illustrates part of an X-ray electrophoto arrangement;

Figure 6 shows a modified arrangement for developing the electric latent image; and

Figures 7 and 8 illustrate another photographic procedure.

My invention contemplates the use of a photoemissive material, that is, material which emits electrons when exposed to radiation such as light, ultra-violet rays, X-rays or the like. According to the invention a layer or surface having photoemissive qualities is exposed to radiation (having a desired image or pattern, for example) to produce a corresponding emission of electrons in the areas receiving the radiation. The electric charge pattern (electrostatic latent image) thus produced on the photoemissive layer or on an adjacent layer of insulation is then developed or rendered visible by depositing a suitable material, such as powder, under influence of the charge pattern.

This may preferably be done automatically immediately after the electric latent image is produced and I have devised certain automatic devices for depositing a finely divided material to develop the image and for fixing the material after deposition. My invention therefore provides an automatic photographing machine comprising a single piece of apparatus for producing a finished print.

While a preferred embodiment of the invention is described herein, it is contemplated that considerable variation may be made in the method of procedure and the construction of parts without departing from the spirit of the invention. In the following description and in the claims, parts will be identified by specific names for convenience, but they are intended to be as generic in their application to similar parts as the art will permit.

Referring to the drawings, Figure 1 shows a device particularly adapted for making photographic reproductions of printed or typewritten matter, drawings, pictures and the like although it may also be used in some cases for general photography. The sheet 10 bearing the printing or drawing to be reproduced is laid on a table and the apparatus is suitably suspended above it with camera portion 11 disposed directly over the sheet 10. Lens 12 is positioned in the lower end of the camera so as to focus the image of the sheet 10 on a photo-electric layer 14 deposited upon a sheet 15 of transparent material supported in the upper part of the camera. Shutter 13 controls the entry of light into the camera. The photoelectric layer 14 is preferably conductive or provided with a contiguous conductor such as a transparent thin metal film deposited upon the sheet 15 before the photoelectric material, or a fine wire screen secured to the surface of sheet 15. A current collecting margin strip 16 of highly conductive material is also provided near the edges of sheet 15 and this is connected to a battery 17 by an insulated conductor 18 and switch 38.

A plane metal electrode 19 is also supported within the camera behind photo-emissive layer 14 and parallel to it. Electrode 19 is connected to the opposite terminal of battery 17 by conductor 20 and switch 37. Battery 17 is a high voltage battery, preferably above 1000 volts or even several thousand volts although the voltage should not be sufficient to start or to maintain a discharge between the surfaces of layers 14 and 19, other than that which is incidental to the photoelectric effect.

A spool or mandrel 21, mounted at the left end of the back part of the camera (as seen in Figure 1) carries a roll of sheet insulating material 22 upon which the photographs are to be produced.

Sheet 22 is of material of high insulating value under the conditions present in the camera and is transparent or translucent. Suitable materials are unplasticized and dehydrated Cellophane, cellulose acetate, translucent paper, transparent sheet resins, polystyrene sheet, paper or Cellophane impregnated with polystyrene, ethyl cel-

Fig. 2.2 Copying by xerography (US 2221776)

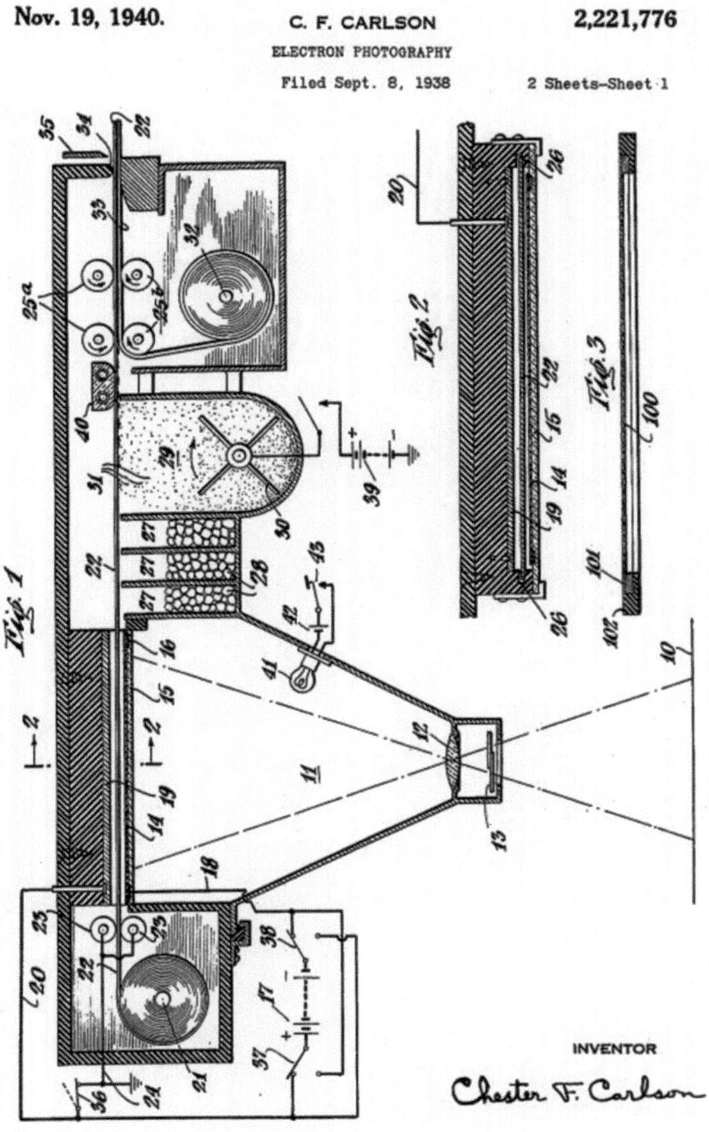

Fig. 2.3 Copying by xerography figures (US 2221776)

Timelines

3

A temporal sequence or a "life cycle" describes the various phases of a patent or utility model. For a patent, the grant of the patent represents a significant milestone. In the period before the grant, the description—not just the set of claims—plays a major role. After the patent is granted, the claims become of paramount importance. Utility models do not exhibit this distinct division, as they do not undergo substantive examination for novelty and inventive step by a patent office.

3.1 German Patent

The beginning of a patent is the filing of the application documents with the patent office. The applicant for a German patent then has up to seven years to file a request for examination.[1] However, if the request for examination is not filed at the same time as the application, or at least promptly, the patent office will not give priority to the examination request, and it cannot be assumed that the first office action assessing the patentability of the invention will be issued before the expiry of the priority period. As a result, when considering whether to file subsequent applications abroad, there is no initial official assessment of patentability to rely on.

The grant procedure before the patent office serves to determine the broadest possible scope of protection. For entering this phase, it is important that the examination procedure begins with a sensibly formulated first set of claims. This can enable a swift grant procedure. It is also important that the description presents the invention comprehensively, in

[1] Section 44 paragraph 2 sentence 1 Patent Act.

T. H. Meitinger, *Patent Made Simple: How to File Without a Lawyer*,
https://doi.org/10.1007/978-3-662-73096-6_3

detail, and concretely. The description should also include different embodiments. It is crucial that the embodiments are described precisely and that it is disclosed how the individual features of the embodiments interact to achieve the inventive effect. This results in a wealth of disclosures that can be used to obtain a legally robust main claim[2] in light of the prior art documents identified by the patent office.

After completion of the grant procedure, the description only plays a secondary role. The description then serves, at most, to aid in understanding the claims.[3] Preferably, the granted patent claims are self-explanatory and define what is protected intellectual property. After the patent is granted, the dependent claims are of only minor importance. Dependent claims serve as fallback positions during the grant procedure. After the end of the grant procedure, they are a relatively superfluous appendage. On the other hand, the dependent claims can also be used to interpret the independent claims. After the patent is granted, essentially only the main claim and the independent claims remain relevant, each of which defines its own scope of protection.

3.2 European Patent

After filing the application documents with the European Patent Office, the filing and search fees must be paid within one month.[4] After payment of the fees, the receiving section checks whether a filing date can be accorded to the application. The application is then examined for formal defects. If the receiving section finds no deficiencies, the search division begins its work and prepares the European search report. This is then sent to the applicant.

After receiving the search report, the applicant has six months to file a request for examination.[5] After the applicant files the request for examination, the examining division examines the application on its merits, in particular for novelty and inventive step. The examining division issues a first communication describing all deficiencies that, at that time, prevent the grant of a patent.[6] The applicant must amend the description and claims accordingly to remedy these deficiencies. This process of substantive examination by the patent division and correction of deficiencies by the applicant is repeated until all

[2] The main claim is the first claim of a set of claims of a patent or utility model. The main claim together with the independent claims define the scope of protection of the patent or utility model.

[3] Section 14 sentence 2 Patent Act or Article 69 paragraph 1 sentence 2 EPC.

[4] Rule 38 paragraph 1 EPC.

[5] Article 94 paragraph 1 sentence 2 in conjunction with Rule 70 paragraph 1 sentence 1 EPC.

[6] Article 94 paragraph 3 EPC.

deficiencies are resolved. Once the examining division has no further objections to granting a patent, the patent is granted.[7] Otherwise, the patent application is refused.[8]

After grant by the European Patent Office, the opposition period begins. Within nine months of publication of the European patent, anyone may file an opposition against the patent. An opposition is only deemed to have been filed once the opposition fee has been paid.[9] In the opposition procedure, the parties first exchange written submissions. Finally, an oral hearing is held before the opposition division. After the oral hearing, the opposition division decides whether the opposition is rejected, the patent is maintained in amended form, or the patent is revoked.[10] If the European patent is maintained in amended form, the European Patent Office publishes a correspondingly amended patent specification.

3.3 International Patent Application

An international patent application[11] can be filed with the international patent office WIPO[12] in Geneva, with the German Patent Office DPMA[13] or with the European Patent Office EPO[14]. At the same time as filing the application documents, a request for search must be filed.[15] A request for examination may also be filed.[16] However, the result of the examination in the international phase is not binding on the patent offices before which the application is continued.

The international phase does not provide for the grant of a patent. Therefore, there is no international patent. Instead, a national grant procedure must be initiated before the patent offices of the countries in which a patent is sought, no later than after 30 or 31 months.[17] An international procedure thus primarily serves to delay national examination procedures. In this way, the costs for national or regional patent grant procedures can be deferred into the future.

---

[7] Article 97 paragraph 1 EPC.

[8] Article 97 paragraph 2 EPC.

[9] Article 99 paragraph 1 sentence 2 EPC.

[10] Article 101 paragraphs 2 and 3 EPC.

[11] Article 3 paragraph 1 PCT.

[12] World Intellectual Property Organization (WIPO), 34, chemin des Colombettes, CH-1211 Geneva 20, Switzerland.

[13] German Patent and Trademark Office (DPMA), Zweibrückenstrasse 12, 80.331 Munich.

[14] European Patent Office (EPO), Bob-van-Benthem-Platz 1, 80.469 Munich.

[15] Article 15 paragraph 1 PCT.

[16] Article 31 paragraph 1 PCT.

[17] In some countries, the national grant procedure must be started no later than after 30 months (examples: China, Canada, and the USA), in other countries no later than after 31 months (examples: South Korea and Russia).

3.4 Utility Model

The division of the life cycle of a patent, which results from the milestone of patent grant, does not occur in the life cycle of a utility model. At most, a cancellation procedure, in which a substantive examination of the utility model for novelty and inventive step is conducted, can be regarded as a comparable turning point. If the utility model survives the cancellation procedure, it can even be considered to have greater legal stability compared to a patent, since the utility model cancellation procedure is at least a bilateral procedure. In the cancellation procedure, the utility model is therefore not only examined by the patent office, but there is also the petitioner of the cancellation procedure, who can submit relevant documents for review and present additional arguments against the legal validity of the utility model.[18]

3.5 Summary

For patents, a distinction can be made between a pre-grant phase and a post-grant phase. Before grant, it is important that the set of claims is suitably drafted to provide a good starting point for the grant procedure. In addition, the description should contain a precise and detailed explanation of the invention and various embodiments of the invention. This allows the description to be used to formulate a patentable main claim if the dependent claims do not provide further assistance. After grant, the independent claims, i.e., the main claim and the auxiliary claims, are of great importance. The description then serves, at most, to provide a better understanding of the claims.[19] For utility models, there is no division of the life cycle, as there is no substantive examination for novelty and inventive step by the patent office. An official examination of legal validity only takes place, if at all, in cancellation proceedings.[20]

A right of priority[21] applies to all intellectual property rights, from which a one-year priority period[22] arises. Within the first year after filing the application documents, a subsequent application for the same invention with the same effective filing date can be filed in any country in the world.[23] Therefore, if it is determined that an industrial property right is also needed in the USA, China, Japan, or a European country, this can be done within the priority period without any problems. A European patent or an international patent application can also be pursued as a subsequent application.

---

[18] Section 15 paragraph 1 Utility Model Act.

[19] Section 14 sentence 2 of the Patent Act or Article 69 paragraph 1 sentence 2 EPC.

[20] Section 15 paragraph 1 of the Utility Model Act.

[21] Article 4A paragraph 1 Paris Convention.

[22] Article 4C paragraph 1 Paris Convention.

[23] An important exception is Taiwan.

A German patent application or a utility model can also be the subsequent application claiming the priority of a foreign application. In this case, the application documents must be filed with the German Patent Office within the one-year priority period of the foreign application. In addition, the Patent Office must be informed that the priority of a foreign application is being claimed. The file number and the country in which the foreign first application was filed must be provided before the expiry of the 16th month after the priority date. In addition, a copy of the earlier application must be submitted within this 16-month period.[24]

[24] Section 41 paragraph 1 sentence 1 of the Patent Act or Section 6 paragraph 2 of the Utility Model Act in conjunction with Section 41 paragraph 1 sentence 1 of the Patent Act.

4 General Formal Requirements

The patent claims, the description, and the abstract must be prepared in a single column as a computer printout. Justified text should not be used. Graphical symbols, chemical or mathematical formulas may be handwritten or drawn if this is not otherwise possible. A line spacing of 1.5 lines must be maintained. For the text of the application documents, for example, Times New Roman (font size 12) or Calibri (font size 11) may be used. Underlining, italics, or boldface should not be used in the text.[1]

DIN A4 paper must be used for the application documents. The paper is to be used in portrait orientation. If necessary, drawings may be made in landscape orientation.[2] The sheets are to be written on one side only. The patent claims, the description, the drawings, and the abstract must each be placed on separate sheets. The sheets of the description are to be consecutively numbered in Arabic numerals (1, 2, 3, etc.) in the footer in the center. Line and paragraph counters must not be used.[3] The header may contain the name of the applicant, the company, or the application reference number.[4] The drawing sheets are to be separately numbered with Arabic numerals, in such a way that first the number of the drawing sheet and then the total number of drawing sheets is indicated. Both numbers are to be separated by a slash (example: 1/6, 2/6, … 6/6).

Minimum margins must be observed on the sheets: Top margin: 2 cm, left margin: 2.5 cm, right margin: 2 cm, and bottom margin: 2 cm.[5] No pictorial representations may be

---

[1] Section 6 paragraph 5 of the Patent Ordinance.

[2] Section 6 paragraph 2 of the Patent Ordinance.

[3] Section 6 paragraph 3 of the Patent Ordinance.

[4] Section 6 paragraph 4 of the Patent Ordinance.

[5] Section 6 paragraph 4 of the Patent Ordinance.

T. H. Meitinger, *Patent Made Simple: How to File Without a Lawyer*,
https://doi.org/10.1007/978-3-662-73096-6_4

included in the continuous text of the description or the claims. An exception applies to chemical or mathematical formulas and to tables.[6]

The individual sheets must not be permanently fastened together (e.g., with eyelets). It is permissible to temporarily fasten the sheets of the application documents, for example with staples or paper clips. Care must be taken to ensure that the fastening leaves at most only minor marks on the edge of the sheets.[7]

[6] Section 5 paragraph 1 sentences 1 and 2 of the Patent Ordinance.

[7] Rule 49 paragraph 4 EPC.

Components of an Application 5

A patent or utility model application comprises several sections that follow a semantic order. The sequence in the application reflects the development history of the invention, with the claims at the end of the application presenting the core of the invention. The claims define the intellectual property being claimed.

In the first section of the application, the introductory part of the description, the state of the art is first explained and the disadvantages arising from it are outlined. This is followed by the technical problem addressed by the invention, namely to provide an advancement of the state of the art in order to overcome its disadvantages. Drawings serve to illustrate specific embodiments of the invention, which are explained in the section "Description of the Drawings" in terms of their features and the interaction of those features. The introductory part of the description and the description of the drawings can be summarized as the "description" of the application documents. A patent application therefore comprises four sections: the drawings, the description, the claims, and the abstract. A utility model application comprises three elements, as an abstract is not required.[1]

The most important part of a granted patent is the claims, especially the main claim. The main claim specifies the subject matter protected as intellectual property. If the set of claims includes dependent claims, these claim further subject matter as intellectual property. In infringement proceedings, it is examined on the basis of the claims whether the potential infringing embodiment can be "read onto" the main claim or one of the dependent claims literally. The description serves at most to provide a better understanding of the claims.[2]

[1] Section 4 paragraph 3 Utility Model Act.

[2] Section 14 sentence 2 Patent Act or Article 69 paragraph 1 sentence 2 EPC.

T. H. Meitinger, *Patent Made Simple: How to File Without a Lawyer*,
https://doi.org/10.1007/978-3-662-73096-6_5

The abstract of the patent application serves solely to inform the public.[3] It is not possible to use features from the abstract to distinguish the main claim or a dependent claim as novel and based on inventive step over the prior art in examination proceedings before the patent office. Therefore, the wording of the abstract should receive the least attention. It is advisable to simply use the wording of the main claim as the abstract.

The claims constitute the most important part of a patent or utility model application. Accordingly, this book provides a detailed explanation of how to draft claims. Numerous practical examples are presented, and recommendations for suitable claim formulations are given based on these examples.

A patent or utility model protects a technical invention. A technical invention is defined by technical features. A claim should therefore be composed exclusively of technical features. Non-technical features, such as descriptions of aesthetic designs like an attractive color pattern that has no technical implications, are disregarded in the examination of protectability by the patent office or in the assessment of an infringement situation by a regular court.

5.1 Introductory Part of the Description

A description, and not just claims, is a mandatory part of an application.[4] The description must describe specific embodiments and provide "support" for the claims.[5] "Support" for the claims is provided if the claims are reproduced verbatim in the application, specifically in the introductory part. The description thus serves to explain the invention in its general form and in specific, particular embodiments. The description should deal exclusively with the invention. Information not necessary for the description of the invention may be omitted.[6] Extensive explanations of the state of the art are not useful.

In the introductory part of the application, the technical field to which the invention belongs is first described in one or two sentences. This is followed by a discussion of the relevant state of the art. This state of the art has disadvantages that lead to the problem addressed by the invention. The very purpose of the invention is to further develop the state of the art and thereby overcome its disadvantages. The claims are then to be presented, and preferably an advantage should be stated for each claim. By stating an advantage for each dependent claim, it becomes more difficult for the examiner in a grant procedure to simply dismiss the dependent claims, which serve as fallback positions, as routine knowledge of a person skilled in the art.

[3] Section 36 paragraph 2 sentence 1 Patent Act or Article 85 EPC.

[4] Article 78 paragraph 1 letter b EPC.

[5] Article 84 sentence 2 EPC.

[6] Section 10 paragraph 3 sentence 1 Patent Ordinance or Rule 48 paragraph 1 letter c EPC.

The description serves in particular for the interpretation of the claims.[7] It is therefore obvious that the description and the claims must not contradict each other, as in such a case interpretation of the claims would not be possible. Interpretation of the claims based on the description would then lead to an understanding of the claims that is in obvious contradiction to the wording of the claims themselves. The description could therefore not serve for interpretation, and the claims themselves would not be clear. The claims, and thus the scope of protection of the patent or utility model, could not be clearly determined. This would result in unclear claims, which would always be interpreted to the detriment of the applicant, since it was within their power to ensure clarity.

Definitions in the description can be useful to clarify terms. However, their binding effect can prove disadvantageous if the definition in the description leads to a contradiction with a claim. Definitions in the description can also be disadvantageous if they result in a broader or narrower definition of a term compared to the claim. A problem therefore arises if the definition in the description is not consistent with the claims. If a term is defined more narrowly in the description than would be understood from the claim alone, the definition disadvantageously leads to a reduction in the scope of protection. Conversely, if the term is defined more broadly in the description than it is understood from the context of the claim, the claim is not sufficiently disclosed. Definitions should therefore be drafted very carefully. It must be ensured that there is no contradiction between definitions in the description and the claims.

5.2 Drawings

It is not mandatory to include drawings with the application. However, the task of the patent application—to describe the invention in detail and comprehensively—is greatly facilitated by drawings. Drawings can be used to illustrate different variants, i.e., embodiments, of the invention from different perspectives (top view, side view, perspective view, cross-sectional view, longitudinal section, and exploded view). A process, for example, can be depicted as a block diagram. The drawings must not be in color. In addition, grayscale shading is not permitted. The drawings must be provided with reference signs, with curved or straight lines drawn from the part of the drawing to be identified to the reference sign. Curved lines have the advantage that they are not interpreted as part of the respective drawing. Drawings can be prepared by specialized patent illustrators. In this case, it can be assumed that there will be no objections due to formal defects in the drawings during the grant or registration procedure.

[7] Section 14 sentence 2 Patent Act or Article 69 paragraph 1 sentence 2 EPC.

5.3 Description of the Drawings

In the application documents, the invention must be explained comprehensively, in all details and possibly in different variants. This can be done in particular by means of a description based on the drawings, describing what is shown in each drawing and how the individual parts interact.

5.4 Claims

The claim defines the intellectual property being claimed. Each claim is described in a single sentence. This sentence contains the features that lead to the realization of the invention. The entirety of the claims is referred to as the set of claims. It is advisable to limit the set of claims to ten claims in the German proceedings, as each additional claim incurs extra costs.[8]

5.5 Public Information: The Abstract

The abstract serves solely to inform the public. In the grant procedure, it is not possible to incorporate features from the abstract into the main claim. Therefore, the least attention should be paid to drafting the abstract. It is sufficient to use the wording of the main claim verbatim as the abstract. A figure must be attached to the abstract. It is advisable to select a figure that contains, as far as possible, the same reference numerals as those used in the claim. The abstract may not exceed 1,500 characters.[9]

[8] Fee item 311 160 in the annex (schedule of fees) to Section 2 paragraph 1 Patent Costs Act (DPMA, https://www.dpma.de/service/gebuehren/patente/index.html, accessed August 6, 2021). In the European procedure, additional claim fees are charged from the 16th claim onwards (EPO, https://my.epoline.org/epoline-portal/classic/epoline.Scheduleoffees, accessed August 6, 2021).

[9] Section 13, Paragraph 1 of the Patent Ordinance.

Introductory Part of the Description 6

The introductory part of an application presents the development history of the invention: what already existed (state of the art), what problems arose, and what objective the inventor therefore set out to achieve. The invention is then presented in its various embodiments, and the advantages of the invention are described.

6.1 Title of the Application

The invention must be given a designation. This designation of the invention constitutes the title of the application. The title serves as the first piece of information to the public about the subject matter of the application. The title of the application must be included in the request for grant. The title should be as concise as possible to illustrate the field of the invention.[1] This facilitates searching for the application. General terms such as "device" or "method" without further specification are not useful. The title may consist of one or more words. Proper names, fanciful terms, the word "patent," or non-technical words that cannot describe the invention must not be used. Abbreviations such as "etc." or "and the like" or "and similar" do not serve for precise description. Trade names and trademarks must not be used. The title of the application is mentioned in the abstract.

[1] Rule 41(2)(b) EPC.

T. H. Meitinger, *Patent Made Simple: How to File Without a Lawyer*,
https://doi.org/10.1007/978-3-662-73096-6_6

6.2 Field of the Invention

The description of the technical field comprises the first two to three sentences of the patent application. Only the state of the art should be described here, i.e., what was already known, and from which technical field the invention originated or to which technical field the invention is to be assigned. The first section of the patent application should also be the shortest section and should address only the essential state of the art with which the invention is concerned.

The function of this "Field of the Invention" section of the patent application is to enable the skilled person to roughly classify the invention within the relevant technical field. In particular, this classification enables the patent office to forward the application to the examining division responsible for the relevant technical field. It is therefore logical that the technical field is determined in accordance with the main claim.

In this section of the application, for example, the generic term of the main claim could be mentioned.[2]

> *The invention relates to a connection point between two tool system parts according to the preamble of claim 1, as well as a tool equipped therewith. The tool system parts in question here may, for example, be a tool head and a holder supporting it, which may be part of a tool extension or, as a tool holder, may be directly part of a machine tool.*[3]

The description of the technical field refers to the state of the art, specifically to the field from which the invention emerged. No features that are new or inventive, and thus form part of the invention, should be explained in this section of the application.

If, in addition to the main claim, there are independent claims, for example a method claim describing how the subject matter of the main claim can be manufactured or used, it may be briefly added that the invention also relates to a method for manufacturing or using the invention.[4]

A single sentence may also suffice to describe the technical field of the invention, for example:

> *"The invention relates to a print nozzle for processing 3D printing material according to the independent claim".*[5]

---

[2] The generic term of a main claim describes the known state of the art from which the invention departs.

[3] DE 10 2019 203888 A1, DPMA, https://depatisnet.dpma.de/DepatisNet/depatisnet?action=pdf&docid=DE102019203888A1&xxxfull=1, accessed on 23 April 2026.

[4] The main claim and the independent claims are the independent claims. The independent claims determine the scope of protection.

[5] DE 10 2019 113993 A1, DPMA, https://depatisnet.dpma.de/DepatisNet/depatisnet?action=pdf&docid=DE102019113993A1&xxxfull=1, accessed on 23 April 2026.

A reference to the independent claims, i.e., the main claim and the independent claims[6] is not strictly necessary. It is also sufficient to state:

> *"The invention relates to a shopping cart for a supermarket and a method for manufacturing such a shopping cart.*

or:

> *The invention lies in the technical field of cutting tools, such as drills or milling cutters. In milling, this feed movement is performed perpendicular or at an angle to the rotational axis of the tool, whereas in drilling, the feed direction is along the rotational axis".*[7]

Another example of a description of the technical field:

> *"The invention relates to a clamping chuck for coupling a tool adapter, in particular an HSK adapter, to a motor spindle of a machine tool, which is suitable for supplying an MQL lubricant (minimum quantity lubrication system) up to a material-removing tool, in particular a rotating drilling, grinding, milling, or countersinking tool".*[8]

This first section of the application can be started with the phrases "The invention relates to…" or "The invention lies in the technical field of …".

6.3 Background of the Invention or State of the Art

In this section, the state of the art must be indicated. The state of the art comprises the documents that can be cited against an invention when it comes to the legal validity of the patent or utility model. Only those documents that were made available to the public before the filing or priority date belong to the state of the art.[9] In this section, not only patent documents, i.e., patents, utility models, patent applications, or utility model

[6] The terms "independent claims" and "dependent claims" originate from the terminology of the EPC, i.e., the European Patent Convention. Independent claims define the scope of protection, while dependent claims select and specify a part of this scope. The main claim is the first claim of a set of claims. The main claim and the independent claims are the independent claims. The terms "main claim," "independent claims," and "subclaims" originate from the terminology of German patent law.

[7] DE 10 2019 109916 A1, DPMA, https://depatisnet.dpma.de/DepatisNet/depatisnet?action=pdf&docid=DE102019109916A1&xxxfull=1, accessed on 23 April 2026.

[8] DE 10 2014 114779 B3, DPMA, https://depatisnet.dpma.de/DepatisNet/depatisnet?action=pdf&docid=DE102014114779B3&xxxfull=1, accessed on 23 April 2026.

[9] Patent documents that were filed with a patent office before the filing or priority date of the relevant intellectual property right and were published only on or after the filing or priority date are only taken into account for the assessment of novelty of the intellectual property right (§ 3(2) sentence 1 Patent Act or Article 54(3) EPC).

applications, but also other written publications may be mentioned. For example, articles from scientific journals may be cited.

For a utility model, a smaller body of prior art is relevant compared to a patent. In utility model law, oral descriptions are not considered prior art. If an invention was told to someone during a train journey, the invention can still be validly protected by a utility model. In addition, public use is only prejudicial to novelty if it took place in Germany.[10] Furthermore, a utility model is subject to a six-month grace period for novelty. The applicant can therefore publish their own invention up to six months before filing the application without this being held against them as relevant prior art.[11]

The applicant is not required to conduct extensive searches or chase after hard-to-obtain documents. The applicant is only expected to indicate the prior art that is currently available to them.[12] "Indicate" according to the Patent Ordinance means that the source must also be specified. For a patent document (filed patent application, filed utility model application, granted patent, or registered utility model), the publication number must be given. For example, one could state: "The document DE102020121932A1 discloses a [title of the invention]." Additionally, parts of the main claim or the abstract could be described. As a rule, one should refrain from formulating an independent interpretation of the technical teaching of the patent document. Providing one's own interpretation of the document always carries the risk that the document itself will be viewed too critically and thus an overly close connection to one's own invention will be constructed. In opposition or nullity proceedings, for example, the opposing party could benefit from this by adopting this line of argument and pointing out that even the inventor themselves saw it this way. It is easy to fall into such a distorting "ex post" perspective. The fact that this risk is significant is evident from the fact that the European Patent Office has developed a special procedure for assessing inventive step to counteract this risk.[13]

The discussion of the prior art should not be too extensive. If this section becomes too lengthy, the applicant's own invention may appear less significant than it actually is.

During the course of the grant procedure, the patent office may require the applicant to include further prior art in the application.[14] This can be done by adding an additional paragraph in this section of the application documents, with the evaluation of the prior art being carried out in the same manner as explained above.

[10] § 3(1) sentence 2 Utility Model Act.

[11] The novelty grace period under utility model law should only be regarded as a last resort in the event of accidental disclosures. Under no circumstances should publication of an invention be planned in reliance on the novelty grace period.

[12] Section 10 paragraph 2 no. 2 of the Patent Ordinance or Rule 42 paragraph 1 letter b EPC.

[13] With the problem-solution approach, the assessment of inventive step is carried out systematically. This approach is now also regularly applied in grant proceedings before the German Patent Office (see 19.6 Problem-Solution Approach).

[14] Section 34 paragraph 7 Patent Act.

An example of an "evaluation of the prior art":

Such hydraulic expansion chucks are known, for example, from WO 2017/093280 A1, DE 102012215036 A1, DE 10312743 A1, DE 102012110392 B4, or WO 2015/166062 A1, and have a base body extending along a rotational or longitudinal central axis, which can functionally be divided into a clamping section for receiving and clamping a shank tool and a shank section.[15]

In this example, reference is first made to the previously described technical field ("Such hydraulic expansion chucks"). The patent documents that were researched must be cited *("WO 2017/093280 A1, DE 102012215036 A1, DE 10312743 A1, DE 102012110392 B4, or WO 2015/166062 A1").* In this way, the applicant already provides the patent examiner with initial documents to conduct the examination of the invention. The applicant thus facilitates the examiner's work and may already indicate a direction for the examination for which the applicant has prepared their application. This may result in a rapid grant of the patent.

Another example of an "evaluation of the prior art":

Generic collet chucks are well known from the prior art and serve as an interface between a machine spindle of a machine tool and a tool, in particular a turning, grinding, drilling, or milling tool. With the tool, which has geometrically defined cutting edges, chips can be removed from a workpiece, with the tool usually rotating. ... For this purpose, for example, EP 16 60 262 B1 and EP 17 13 606 B1 disclose a collet chuck in which a ...[16]

The phrase "Generic ... are well known from the prior art, ..." can be omitted, as the inventor naturally assumes that the features of the preamble are known.

Preferably, patent documents[17] should be cited in the prior art. It is not sufficient to describe an alleged prior art without reference to patent documents. This section only becomes effective if patent documents are cited. Only in this case will the examiner seriously consider this section and use it as a starting point for their own search.

Further examples of an "evaluation of the prior art":

DE 10 2011 106 421 B3 discloses a hydraulic expansion device that teaches a hydraulic expansion chuck with a clamping sleeve. To limit movement of the tool shank in the clamping sleeve, a locking element is inserted into the clamping sleeve as a releasable tool securing device.[18]

---

[15] DE 10 2019 209732 A1, DPMA, https://depatisnet.dpma.de/DepatisNet/depatisnet?action=pdf&docid=DE102019209732A1&xxxfull=1, accessed on 23 April 2026.

[16] DE 10 2014 114779 B3, DPMA, https://depatisnet.dpma.de/DepatisNet/depatisnet?action=pdf&docid=DE102014114779B3&xxxfull=1, accessed on 23 April 2026.

[17] Patent documents can be granted patents, filed patent applications, filed utility model applications, or registered utility models. It does not matter from which country the patent document originates or what its grant status is.

[18] DE 10 2014 101122 B3, DPMA, https://depatisnet.dpma.de/DepatisNet/depatisnet?action=pdf&docid=DE102014101122B3&xxxfull=1, accessed on 23 April 2026.

"There are essentially two types of coffee machines on the market: so-called 'semi-automatic' or 'portafilter machines,' in which the ground coffee from a grinder placed next to the coffee machine is dosed into a portafilter (brewing pan with brew sieve and outlet opening), which is then clamped into the coffee machine (see, for example, publication EP 2 314 188 A1). The brewing chamber is then ..."[19]

As prior art, for example, DE 149 1611 A discloses a portable heart massage device comprising a frame, a stand, and a base plate with an arm projecting freely above the base plate, which carries a massage cylinder. The device relates to cardiac massage in the event of a heart attack and, despite its portability, is not a handheld device and for these reasons alone is not suitable for home use.[20]

The formulation "As prior art, for example, ... discloses a ... comprising/with/which includes ..." is elegant. Alternatively, a formulation such as "The document DE... shows/describes/discloses ..." is also common.

6.4 Summary of the Invention

From this section onward, the description of the invention begins. Previously, the prior art was mainly explained.

6.4.1 Disadvantages of the Prior Art

In this section, the disadvantages of the prior art are to be elaborated. These disadvantages directly give rise to the object of the invention, since the object of the invention is precisely to overcome the disadvantages of the prior art. Therefore, the disadvantages of the prior art should be used to lead into the object of the invention.

Describing the disadvantages:

"However, this design requires skilled personnel who are proficient in adjusting the grind size and tamping—otherwise, highly variable coffee quality is to be expected."[21]

"The coffee quality of such machines, even with optimal design and optimal settings, is not as good as that of a semi-automatic."[22]

[19] WO 2013117362 A1, DPMA, https://depatisnet.dpma.de/DepatisNet/depatisnet?action=pdf&docid=WO002013117362A1&xxxfull=1, accessed on 23 April 2026.

[20] DE 20 2018 103631 U1, DPMA, https://depatisnet.dpma.de/DepatisNet/depatisnet?action=pdf&docid=DE202018103631U1&xxxfull=1, accessed on 23 April 2026.

[21] WO 2013117362 A1, DPMA, https://depatisnet.dpma.de/DepatisNet/depatisnet?action=pdf&docid=WO002013117362A1&xxxfull=1, accessed on 23 April 2026.

[22] WO 2013117362 A1, DPMA, https://depatisnet.dpma.de/DepatisNet/depatisnet?action=pdf&docid=WO002013117362A1&xxxfull=1, accessed on 23 April 2026.

6.4.2 Objective of the Invention

The technical problem to be solved must be stated in the application. This is particularly required if specifying the technical problem is "indispensable for understanding the invention or for its more precise substantive definition."[23] The objective of the invention is thus the solution to the technical problem. When formulating the objective, the following point should be observed:

Tip: The formulation of the objective should make the technical character apparent.

An invention is the solution to a technical problem by technical means.[24] Accordingly, it should be clear that the invention addresses a technical objective. General formulations that do not reveal the technical character of the objective should be avoided. For example, a formulation such as "to provide a device" or "to enable" is not necessarily technical. Where possible, a more specific formulation is preferable, for example "to improve braking performance," "to increase ease of maintenance," or "to avoid rejects." Other specific objectives could be "to reduce manufacturing costs" or "to enable better handling by a user." General formulations of objectives do not help to substantiate the patentability of an invention. To "improve" or "optimize" an object is not a specific technical objective.[25]

Tip: The formulation of the objective should be specific.

A formulation of the objective such as "to arrange the inventive features X, Y, and Z on the device" would be disadvantageous, because in this way the features X, Y, and Z would no longer be part of the invention, but only their use. This would diminish the inventive step. Therefore, the objective should not contain any elements of the solution to the invention. The objective of the invention should not already indicate the solution path. After reading the objective, the "how is this achieved" should not be obvious.

Tip: The objective must not anticipate features of the invention.

Examples from practice:

> *"The present invention is therefore based on the objective of, during the operation of a chip-removing tool with minimum quantity lubrication, shortening the time delay in supplying the working area of the tool with a sufficient amount of cooling and lubricating medium, and*

---

[23] Section 10 paragraph 2 no. 3 Patent Ordinance.

[24] Section 1 paragraph 1 Patent Act.

[25] Schulte/Moufang, Patent Act with EPC, Commentary, 10th edition, Section 1 Patentable Inventions, margin no. 50.

making the cooling and/or lubrication of the working area or the tool cutting edge(s) more uniform."[26]

or:

"It is therefore an objective of the invention to provide a coffee machine that combines the advantages of both types of construction, namely the simplicity and high possible coffee quality of the semi-automatic and the consistency of results of the fully automatic. If possible, the workload for the operator should also be reduced compared to the semi-automatic. It is a further objective of the invention to specify a method for operating such a coffee machine."[27]

or:

"The objective of the invention is to provide a burnishing tool that enables good machining of workpieces while maintaining a high machining speed."[28]

In document DE 10 2019 109692 A1, the objective is described as:

"The objective of the present invention is to overcome the disadvantages of the prior art and, in particular, to provide a milling tool that is suitable for milling fiber-reinforced materials, especially in the case of complex component geometries with different fiber types."[29]

In document DE 10 2014 106968 A1, the objective is described as:

"The objective of the invention is to design an extraction device for a machine tool and a machine tool of the type mentioned at the outset, with which the removal of abraded material can be carried out as simply and efficiently as possible. In addition, the extraction device should be as space-saving as possible, especially in the area of at least one working section. The extraction device should be as easy to operate as possible."[30]

The objective can therefore generally be described as follows:

The objective of the invention is to provide a device and a method so that efficient processing/extraction/production/assembly/conversion/retrofitting of a machine/tool/device/vehicle can be achieved. In addition, the device should be compact/space-saving/cost-effective to manufacture/easy to assemble/produce/handle/operate.

---

[26] DE 10 2012 224287 A1, DPMA, https://depatisnet.dpma.de/DepatisNet/depatisnet?action=pdf&docid=DE102012224287A1&xxxfull=1, accessed on 23 April 2026.

[27] WO 2013117362 A1, DPMA, https://depatisnet.dpma.de/DepatisNet/depatisnet?action=pdf&docid=WO002013117362A1&xxxfull=1, accessed on 23 April 2026.

[28] DE 10 2019 111784 A1, DPMA, https://depatisnet.dpma.de/DepatisNet/depatisnet?action=pdf&docid=DE102019111784A1&xxxfull=1, accessed on 23 April 2026.

[29] DE 10 2019 109692 A1, https://depatisnet.dpma.de/DepatisNet/depatisnet?action=pdf&docid=DE102019109692A1&xxxfull=1, accessed on 23 April 2026.

[30] DE 10 2014 106968 A1, https://depatisnet.dpma.de/DepatisNet/depatisnet?action=pdf&docid=DE102014106968A1&xxxfull=1, accessed on 23 April 2026.

The objective should not be formulated too abstractly, since the invention is intended to solve a specific technical problem. Thus, an objective such as "to enable technical progress" would be unsuitable. On the other hand, the objective should not be described in too much detail, otherwise doubts may arise as to the sufficiency of the inventive step. For example, an objective to modify a tool in such a way that it has a threaded bolt would lead directly to the invention, namely to provide a tool with a threaded bolt.

It is possible to amend the objective during the course of an examination procedure. This may be necessary in particular if prior art documents become known that render the solution to the previous objective no longer novel or inventive.

6.4.3 Support of the Claims in the Description

This section of the application can be introduced with the following passage, which provides a transition from the technical problem to the invention: "The problem is solved by the features of the independent claim or claims. Advantageous developments of the invention are described in the dependent claims."

In this section, the claims are to be repeated verbatim. The independent claims can be introduced with the phrases: "According to a first aspect, the invention according to the invention is provided by [main claim]." or "According to a first aspect, a [main claim] is provided." or "A first aspect of the invention is [main claim]." For the further independent claims, the following formulations can be used: "According to a second/third/fourth aspect, the invention according to the invention is provided by [further independent claim].", "According to a second/third/fourth aspect, a [further claim] is provided." or "A first/second/third/fourth aspect of the invention is [further claim]."

After the independent claims, the dependent claims should also be included in the introductory part of the application. This can be done, for example, using the following formulations:

> "Exemplary embodiments are described in the dependent claims.
> According to an exemplary embodiment of the invention, a [dependent claim 1] is provided.
> According to a further exemplary embodiment of the present invention, a [dependent claim 2] is provided.
> In a further embodiment according to the invention, a [dependent claim 3] is provided."

A special variation of the introductory wording is not required. Instead of reproducing the dependent claims verbatim, a formulation such as "Advantageous embodiments are the subject of the dependent claims." could be used. However, it is recommended to actually repeat the dependent claims verbatim. This allows the advantages or beneficial effects of each dependent claim to be described.[31] This preempts a possible argument in

[31] Section 10 paragraph 2 no. 6 Patent Ordinance or Rule 42 paragraph 1 letter c EPC.

examination proceedings that the respective dependent claim merely reflects the routine skill of the average person skilled in the art (and therefore the features of the dependent claim cannot establish inventive step). If no advantageous effects can be identified for a particular dependent claim, one should reconsider whether it actually represents a meaningful fallback position and should therefore be included in the set of claims. It is not advisable to include features in dependent claims that, in combination with the referenced main claim (or referenced further independent claim), cannot lead to an allowable claim. The chance of a rapid grant of the patent could otherwise be lost.

Furthermore, responding to an official communication from the examining division is significantly facilitated if the advantageous effects of the respective dependent claims have already been described. This allows the problem-solution approach to be prepared almost automatically.[32] It is also advisable, in addition to the advantages of the subject matter of the dependent claims, to include explanations and definitions of the features of the dependent claims. This allows clarifications or alternative implementations of the features to be explained.

In this section, no reference signs should be used when reproducing the claims. The reference signs are only introduced in the section "Brief Description of the Drawings." In the description (sections "Brief Description of the Drawings" and "Detailed Description of Exemplary Embodiments"), the reference signs are used without parentheses. In the claims, the reference signs are set in parentheses.

In the utility model specification DE 20 2019 000855 U1, the following passage can be found: "Three advantageous embodiments of the invention are described in claims 5, 6, and 7 as well as in Fig. 2, Fig. 4, and Fig. 5."[33] Before and after this, the utility model does not address the subject matter of the other claims. This passage therefore states that only the subject matter of claims 5, 6, and 7 are advantageous, but not the other claims. For the applicant, this passage is therefore disadvantageous, as only claims 5, 6, and 7 are presented as relevant. However, arguments should be provided for each claim in the set of claims to justify that the claim is considered protectable. In addition, all claims in the application should generally be designated as being based on an inventive step.

[32] See 19.6 Problem-Solution Approach, guidelines of the EPO, https://link.epo.org/web/legal/guidelines-epc/en-epc-guidelines-2026-hyperlinked.pdf, accessed on 23 April 2026.

[33] Section [0003] of DE 20 2019 000855 U1, DPMA, https://depatisnet.dpma.de/DepatisNet/depatisnet?action=pdf&docid=DE202019000855U1&xxxfull=1, accessed on 23 April 2026.

7 Drawings

Drawings are the language of the engineer. As a rule, it is advisable to include drawings with an application. In particular, drawings can be used to describe different embodiments of the invention. Furthermore, the invention can be illustrated more effectively through drawings in various perspectives (side view, top view, perspective view, and exploded view). These do not necessarily have to be computer-generated drawings. Freehand drawings may be sufficient. If the patent office objects to the drawings, drawings in compliance with patent requirements can be submitted later.

Text passages in the drawings are not permitted. However, individual words may be used to aid understanding, for example, "voltage," "current," "pressure," or "air." The relevant elements in the drawings must be identified with reference numerals (1, 2, 3, 4, …), with the reference numerals placed at the edge of the drawings and lines or arrows leading from the numerals to the elements to be identified in each drawing. The reference numerals may be used in the description (not in parentheses) and in the claims (in parentheses). The drawings must be submitted together with the other documents of the application.

The reference numerals establish a clear relationship between the description, the claims, and the drawings. If several drawings are used, the same elements should be identified with the same reference numerals.

The Patent Ordinance[1] must be observed when preparing the drawings. In particular, Annex 2 to § 12 of the Patent Ordinance (Standards for the Submission of Drawings) with items 2, 3, 5, and 7 must be taken into account.

[1] Patent Ordinance of September 1, 2003 (Federal Law Gazette I p. 1702), as last amended by Article 1 of the Ordinance of December 12, 2018 (Federal Law Gazette I p. 2446).

T. H. Meitinger, *Patent Made Simple: How to File Without a Lawyer*,
https://doi.org/10.1007/978-3-662-73096-6_7

Annex 2, item 2 to § 12 of the Patent Ordinance:

"The drawings must be executed with sufficient contrast, in permanent, black, sufficiently strong and dark, uniformly consistent and sharply defined lines and strokes, without the use of color."

Annex 2, item 3 to § 12 of the Patent Ordinance:

"In addition to views and sectional drawings, perspective views or exploded views may also be used to illustrate the invention. Cross-sections must be indicated by hatching, which must not impair the legibility of the reference numerals and leader lines."

Annex 2, item 5 to § 12 of the Patent Ordinance:

"The lines in the drawings should not be drawn freehand, but with drawing instruments. The numerals and letters used in the drawings must be at least 0.32 cm high. Latin letters and, where customary, Greek letters are to be used for labeling the drawings."

Annex 2, item 7 to § 12 of the Patent Ordinance:

"Reference numerals may only be used in the drawings insofar as they are listed in the description and, where applicable, in the patent claims, and vice versa. The same applies to the abstract and its drawing."

Annex 2, item 8 to § 12 of the Patent Ordinance:

"The drawings must not contain any explanations; exceptions are short, indispensable indications such as 'water,' 'steam,' 'open,' 'closed,' 'section along A-B,' as well as, in electrical circuit diagrams and block diagrams or flowcharts, short keywords that are indispensable for understanding."

7.1 Use of Reference Numerals

In the upper figure of Fig. 7.1, reference numeral 1 identifies both the rectangle and the circle. In the middle figure, reference numeral 1 identifies only the circle within the rectangle, and in the lower figure of Fig. 7.1, reference numeral 1 identifies the upper surface (upper with respect to the drawing plane) of the rectangle.

Instead of numbers, letters can also be used as reference numerals. However, it has become customary to use Arabic numerals and to use letters only if there is a particular reason to do so.

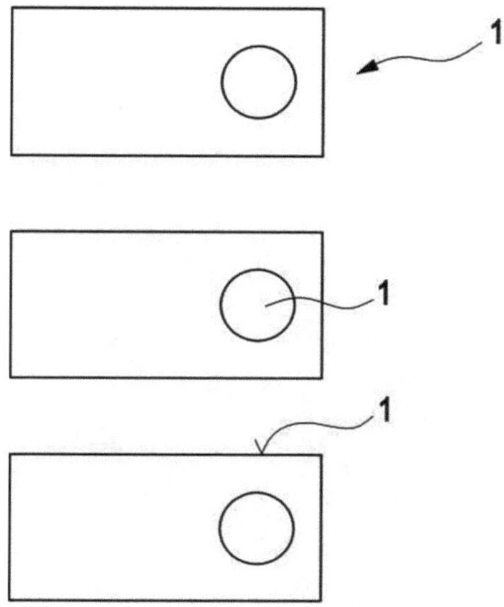

Fig. 7.1 Use of reference numerals

7.2 Recommended Practical Examples

The DE 10 2019 111784 A1 shows the following drawing (see Fig. 7.2, DPMA, https://depatisnet.dpma.de/DepatisNet/depatisnet?action=pdf&docid=DE102019111784A1&xxxfull=1, accessed on 23 April 2026).

The DE 10 2019 109 692 A1 subsequently shows the following drawing (see Fig. 7.3, DPMA, https://depatisnet.dpma.de/DepatisNet/depatisnet?action=pdf&docid=DE102019109692A1&xxxfull=1, accessed on 23 April 2026).

In the document DE 10 2014 106 968 A1, the following drawing is shown (see Fig. 7.4, DPMA, https://depatisnet.dpma.de/DepatisNet/depatisnet?action=pdf&docid=DE102014106968A1&xxxfull=1, accessed on 23 April 2026).

It can be useful to represent a method, for example one included as a method claim in the set of claims, by means of a flowchart or block diagram. In WO2013117362A1, a flowchart is shown (see Fig. 7.5, DPMA, https://depatisnet.dpma.de/DepatisNet/depatisnet?action=pdf&docid=WO002013117362A1&xxxfull=1, accessed on 23 April 2026).

Although a drawing should not contain any text, it has proven useful, especially in flowcharts, to include at least individual words or short text passages in order to maintain comprehensibility.

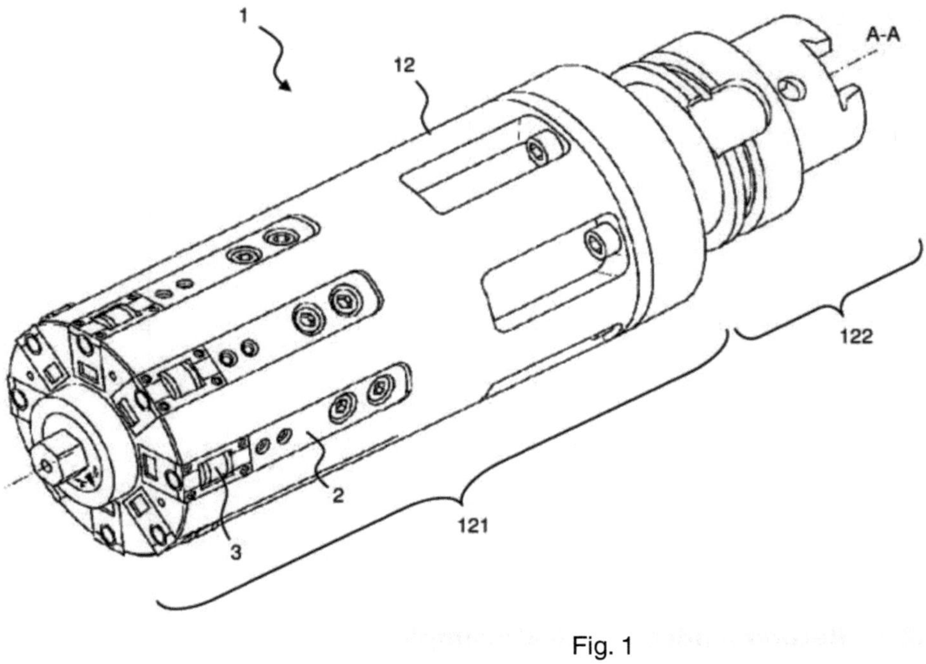

Fig. 7.2 Burnishing tool (DE 10 2019 111 784 A1)

Photographs are only permitted if a drawn solution is not possible. It should be borne in mind that photographs are only reproduced and filed in black and white. If important details can only be recognized by color differences, a color photograph does not provide a suitable means of representation.[2] In such cases, hatching may be an option to illustrate the technical teaching.

[2] EPO, Guidelines for Examination, Part A, Chapter IX-1, Section 1.2 (retrieved on 23 May 2021, https://www.epo.org/law-practice/legal-texts/html/guidelines/d/a_ix_1_2.htm).

Fig. 7.3 Milling tool (DE 10 2019 109 692 A1)

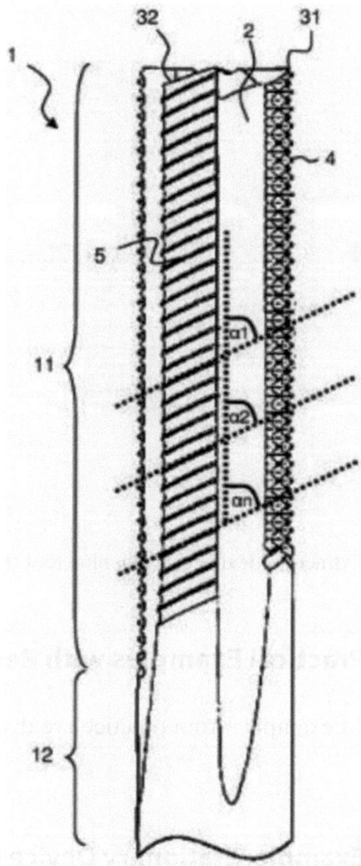

Fig. 1

In a German or European patent application, the drawings must be arranged on separate sheets. Each drawing must be labeled consecutively (Fig. 1, Fig. 2, etc.). In addition, the pages on which the drawings are shown must be numbered (for example, with 10 pages of drawings: 1/10, 2/10, …, 10/10). It is not permitted to include drawings in the running text of the description.[3]

[3] Rule 49, paragraph 9, sentence 1 EPC.

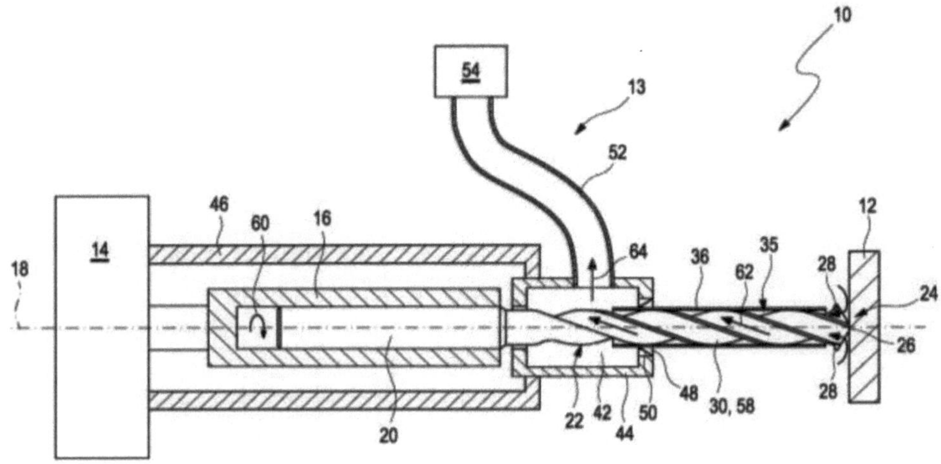

Fig. 1

Fig. 7.4 Extraction device of a machine tool (DE 10 2014 106 968 A1)

7.3 Practical Examples with Recommendations

Real-world examples from practice are discussed, for which improved drawings are presented.

7.3.1 Example: Stationary Device for Securing a Bicycle Against Theft

In the document DE 19610721 A1 (Title: Stationary device for securing a bicycle against theft), the following drawing is presented (see Fig. 7.6, DPMA, https://depatisnet.dpma.de/DepatisNet/depatisnet?action=pdf&docid=DE000019610721A1&xxxfull=1, accessed on 23 April 2026).

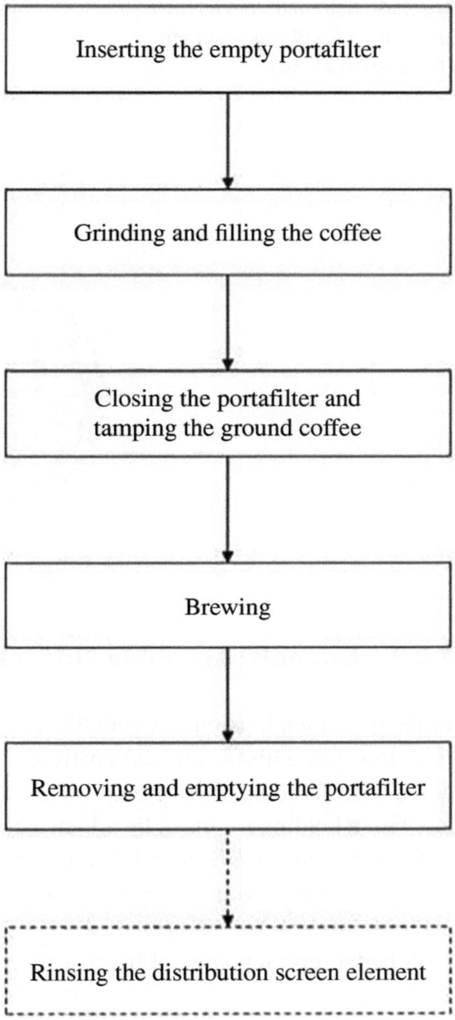

Fig. 7.5 Coffee machine flowchart (WO 2013117362 A1)

There are black areas in this drawing. This is not permitted. At most, cross lines may be inserted into an area to indicate that it is a sectional view. One reason for this is that the lines of the reference numerals can no longer be recognized if they pass through a black area.

A better representation of the invention is shown in Fig. 7.7.

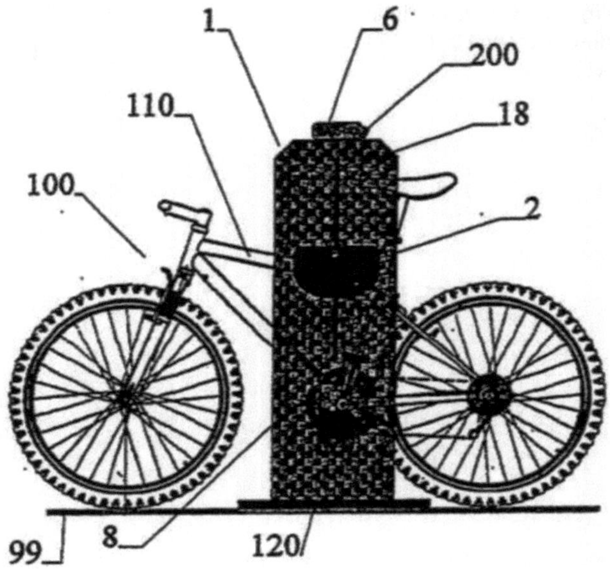

Fig. 7.6 Stationary device for securing a bicycle against theft (DE 19610721 A1)

7.3.2 Example: Assembly Aid for a Bicycle

In the patent application DE 19609598 A1 (Title: Assembly aid for a bicycle), a drawing (see Fig. 7.8, DPMA, https://depatisnet.dpma.de/DepatisNet/depatisnet?action=pdf&docid=DE000019609598A1&xxxfull=1, accessed on 23 April 2026) shows an assembly aid with a holding means 6, by which a bicycle 2 is held, namely above the ground, so to speak in the air. This is intended to facilitate repairs to the bicycle. This drawing contains black areas (for example, the pedals of the bicycle 2) and gray-shaded areas. Black areas and gray shading are not permitted.

A better representation of the invention is shown in Fig. 7.9.

7.3.3 Example: Automatic Control of Functions on Bicycles

The patent specification DE 10 2013 013 406 B4 (Title: Device for the automatic control of the height adjustment of a saddle and other functions on a bicycle with and without motor assistance) includes two drawings (see Fig. 7.10, DPMA, https://depatisnet.dpma.de/DepatisNet/depatisnet?action=pdf&docid=DE102013013406B4&xxxfull=1, accessed on 23 April 2026).

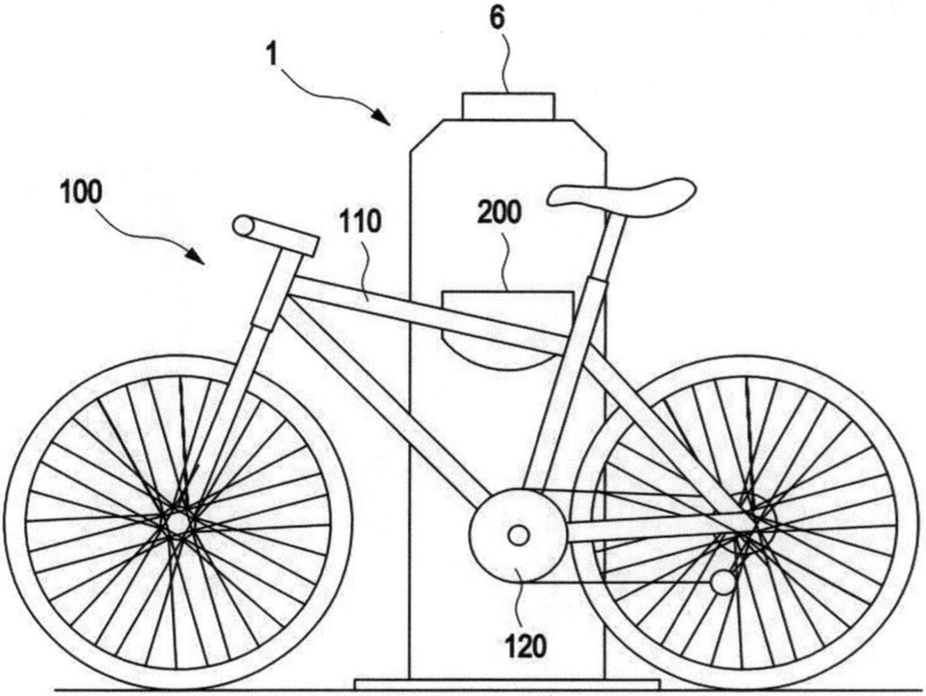

Fig. 7.7 Stationary device for securing a bicycle against theft (DE 19610721 A1)—corrected

The drawings are freehand sketches that were accepted as such by the patent office. It is not required to create computer-generated drawings. However, the drawings are not entirely clear. Some reference numerals and their leader lines are difficult to discern. Any lack of clarity in the disclosure is always to the detriment of the applicant. Great importance should be attached to clear and comprehensible drawings.

7.3.4 Example: Drinking Aid

The utility model specification DE 20 2008 000 979 U1 (Title: Drinking Aid) contains the following drawing (see Fig. 7.11, DPMA, https://depatisnet.dpma.de/DepatisNet/depatisnet?action=pdf&docid=DE202008000979U1&xxxfull=1, accessed on 23 April 2026).

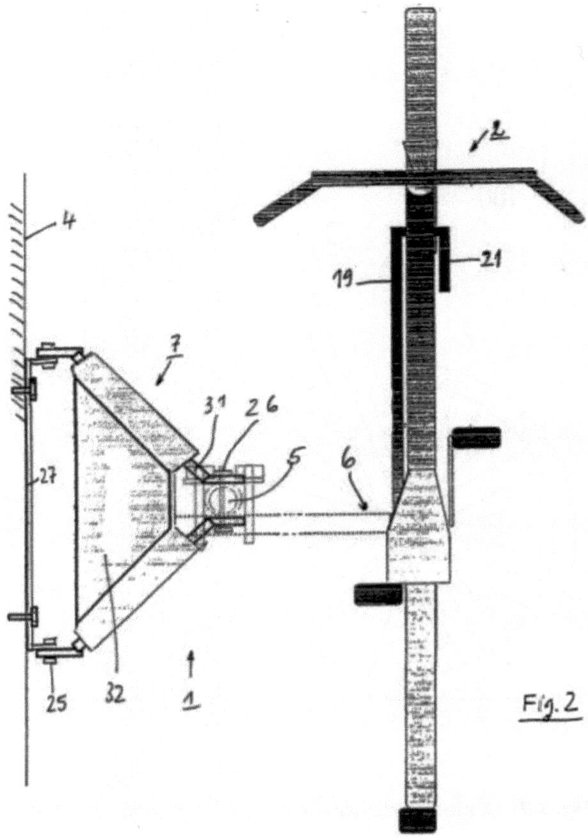

Fig. 7.8 Assembly aid for a bicycle (DE 19609598 A1)

The figure labels, i.e., Fig. 1, Fig. 2, etc., should not be boxed as done here. This drawing shows a drinking cup from above, with all elements of the cup arranged in duplicate, i.e., opposite each other. In this case, it is useful to additionally label the same elements (with the same) reference numerals. This makes it easier to identify and distinguish the individual elements. Below is the corrected drawing with additional reference numerals (see Fig. 7.12).

7.3.5 Example: Baby Bottle Sponge Attachment

The utility model DE 20 2021 000 115 U1 (Title: Drinking Aid, Baby Bottle Sponge Attachment, Bib Substitute, Bib Holder, Pacifier Aid) includes in particular Figs. 7 and 8 (see Fig. 7.13, DPMA, https://depatisnet.dpma.de/DepatisNet/depatisnet?action=pdf&docid=DE202021000115U1&xxxfull=1, accessed on 23 April 2026).

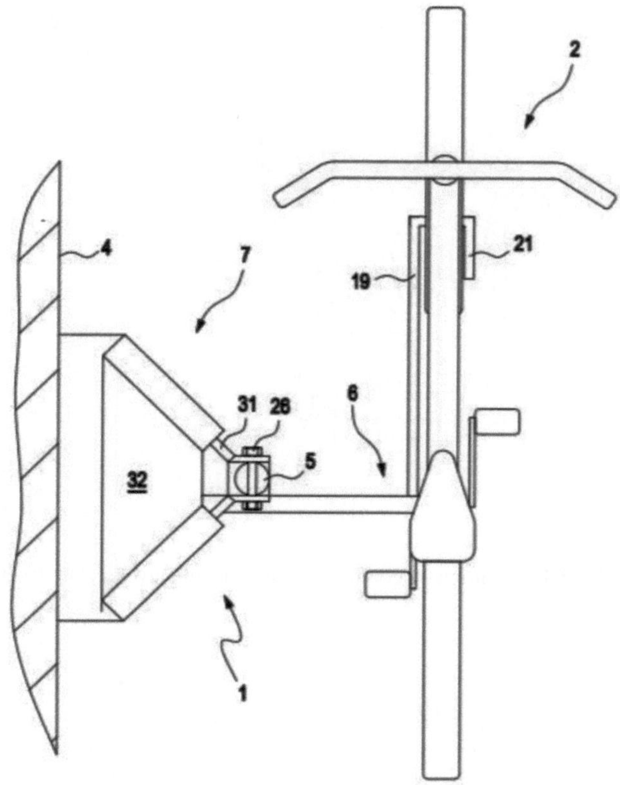

Fig. 1

Fig. 7.9 Assembly aid for a bicycle (DE 19609598 A1)—corrected

It would have been better to use curved lines to indicate the parts of the drawings that are identified by the reference numerals.

A reference numeral list is also included (see Table 7.1).

First, it can be noted that the designation of the reference numerals for Fig. 8 is correct, but not for Fig. 7. Reference numerals must be used consistently for all drawings in a patent or utility model application. Here, reference numeral "1" in Fig. 8 denotes the teat of the baby bottle, as indicated in the reference numeral list. In Fig. 7, however, reference numeral "1" does not denote the teat of the baby bottle, but the baby bottle itself.

Furthermore, in the drawings, a period is placed after the reference numerals. A period is not necessary. If a reference numeral "4.1" is used, a period can serve as a separator. Alternatively, the reference numeral "41" could be used. However, it is neither necessary nor desired to place a period after every reference numeral as a terminator.

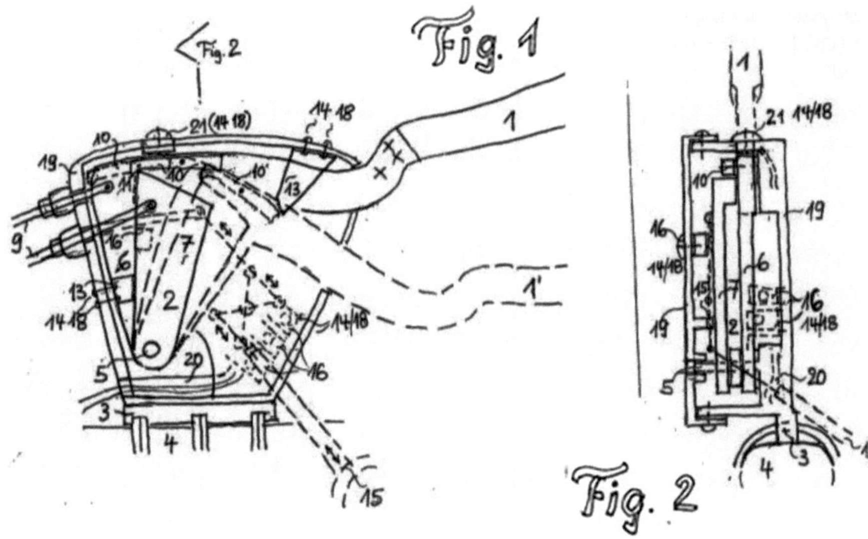

Fig. 7.10 Automatic control of functions on bicycles (DE 10 2013 013 406 B4)

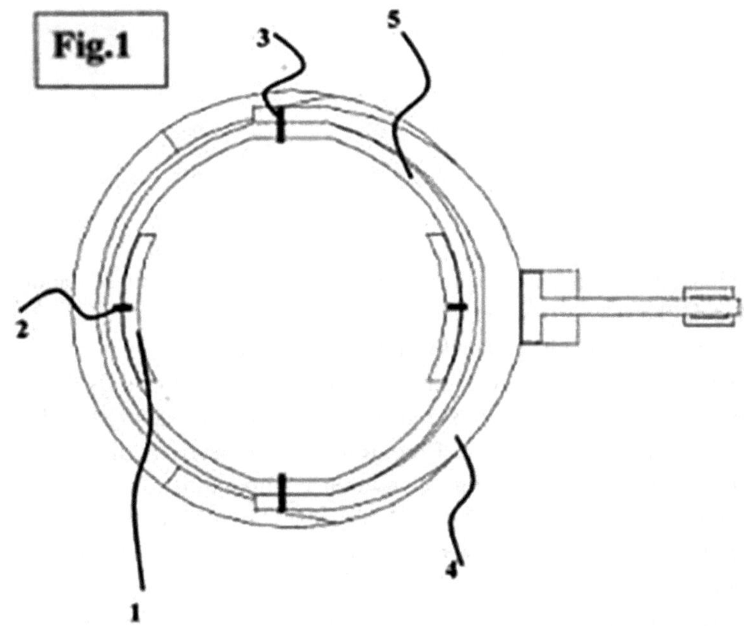

Fig. 7.11 Drinking aid (DE 20 2008 000 979 U1)

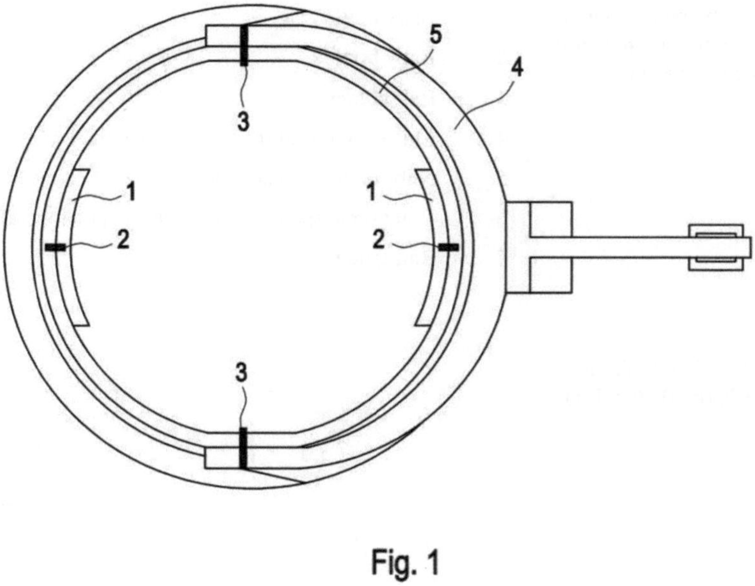

Fig. 7.12 Drinking aid (DE 20 2008 000979 U1)—corrected

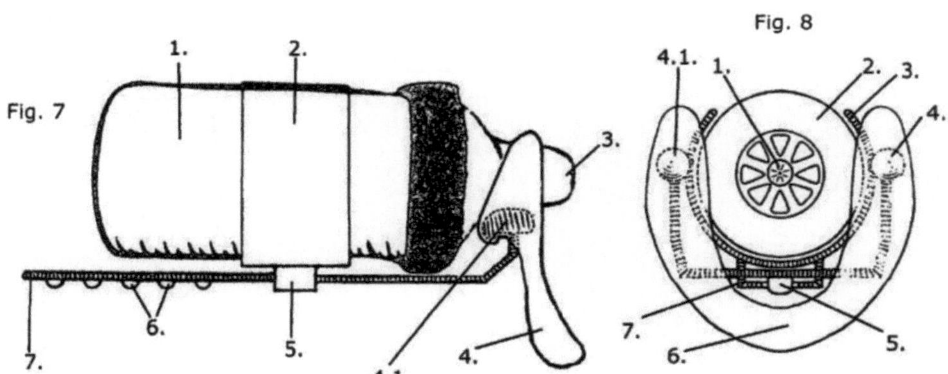

Fig. 7.13 Baby bottle sponge attachment (DE 20 2021 000115 U1)

7.3.6 Example: Hot Dog Drinking Cup

The utility model specification DE 20 2016 002 669 U1 (Title: Hot Dog (Curry) Sausage Croissants Drinking Cup) includes the following three drawings with a reference numeral list (see Fig. 7.14 and Table 7.2, DPMA, https://depatisnet.dpma.de/DepatisNet/depatisnet?action=pdf&docid=DE202016002669U1&xxxfull=1, accessed on 23 April 2026).

Table 7.1 Baby bottle sponge attachment—reference numeral list (DE 20 2021 000 115 U1)

1	Baby bottle teat
2	Baby bottle
3	Rail holder
4	Mushroom head for sponge attachment
5	Nubs on the adjustment aid
6	Ergonomic sponge attachment shaped towards the baby
7	Rail guide

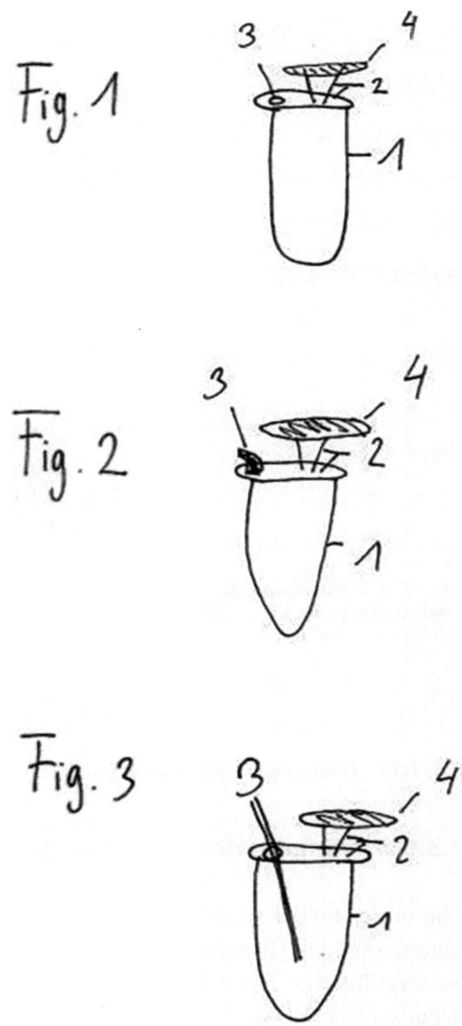

Fig. 7.14 Hot dog drinking cup (DE 20 2016 002 669 U1)

Table 7.2 Hot dog drinking cup—reference numeral list (DE 20 2016 002 669 U1)

Reference list	
1	Holder for sausage, hot dog, currywurst, or croissant
2	The attachment lid
3	Opening for e.g. drinking straw
4	Sausage or croissant
5	(plastic, cardboard) drinking cup made of

The utility model specification contains Figs. 1 to 3, which are executed as freehand drawings. It is permissible to submit drawings as freehand sketches. However, care should be taken to ensure that the same element is always identified with the same reference numeral. For example, in Fig. 1, reference numeral 3 denotes an opening for, for example, a drinking straw. In Fig. 2, reference numeral 3 denotes a drinking aid, and in Fig. 3, the impression is given that reference numeral 3 refers to a drinking straw. Furthermore, it is not clear what is meant by reference numeral 2.

7.3.7 Example: Flexible Mount for a Children's Toy

Utility model DE 20 2007 001 551 U1 (Title: Flexible Mount for Children's Toy, which is mounted on strollers) addresses the technical problem of providing a toy holder that can be mounted on a stroller. The toy holder is particularly characterized by not restricting the freedom of movement of the children (see Figs. 7.15 and 7.16, DPMA, https://depatisnet.dpma.de/DepatisNet/depatisnet?action=pdf&docid=DE202007001551U1&xxxfull=1, accessed on 23 April 2026).

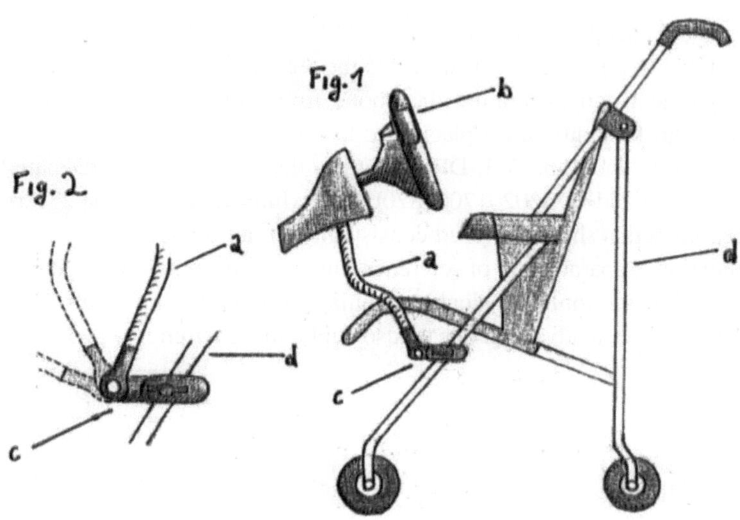

Fig. 7.15 Flexible mount for a children's toy Figs. 1 and 2 (DE 20 2007 001 551 U1)

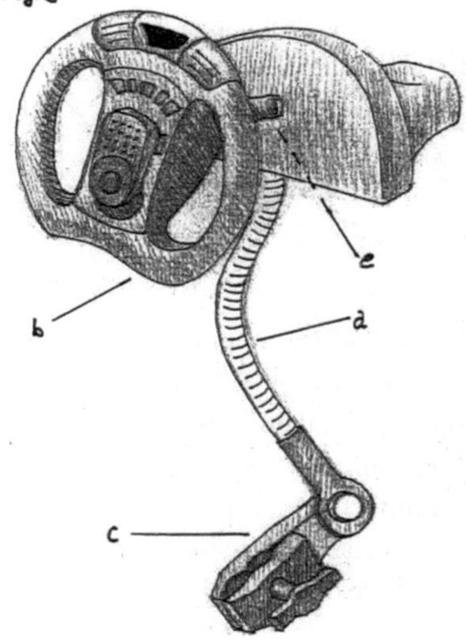

Fig. 7.16 Flexible mount for a children's toy Fig. 3 (DE 20 2007 001 551 U1)

The utility model specification includes three drawings. These drawings show hatched areas, which create a three-dimensional or 3D effect. Such hatching should be avoided. Hatching may only be used in a sectional drawing to indicate that it is a section (longitudinal or cross-section).

A better representation of the invention is shown in 7.17, 7.18, and 7.19.

7.3.8 Example: Foldable Mobile Phone Case on Textiles

The utility model specification DE 20 2017 001 820 U1 (Title: Foldable mobile phone case (smartphone case) on textiles) addresses the technical problem of carrying a mobile phone in a garment and preventing the phone from falling out when being removed from the garment. In addition, misplacing or losing the phone should be prevented (see Figs. 7.20 and 7.21 and Table 7.3, DPMA, https://depatisnet.dpma.de/DepatisNet/depatisnet?action=pdf&docid=DE202017001820U1&xxxfull=1, accessed on 23 April 2026).

Reference numerals should be used consistently for all drawings in a patent or utility model application. Separate lists of reference numerals for individual figures should not be created. This avoids confusing double meanings of reference numerals. In these drawings, it would have been advantageous to also depict the smartphone.

Fig. 7.17 Flexible mount for a children's toy Fig. 1 (DE 20 2007 001 551 U1)—corrected

1/2

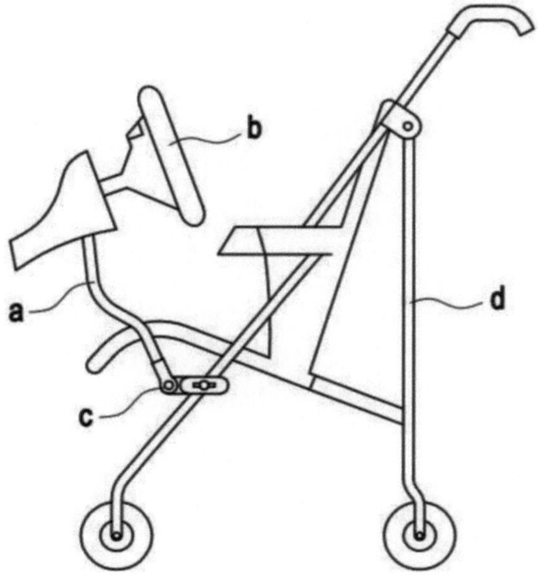

Fig. 1

Fig. 7.18 Flexible mount for a children's toy Fig. 2 (DE 20 2007 001 551 U1)—corrected

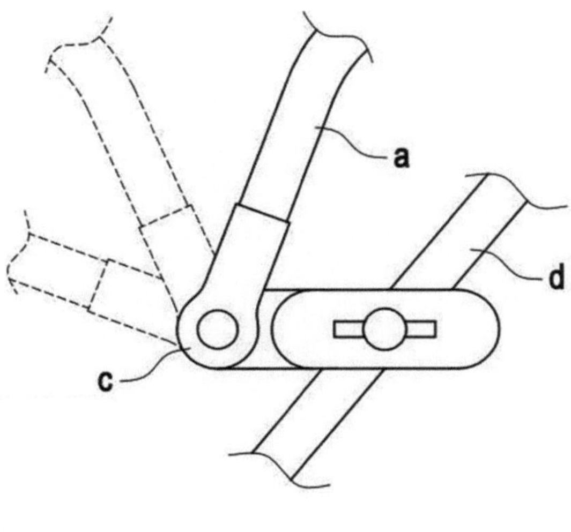

Fig. 2

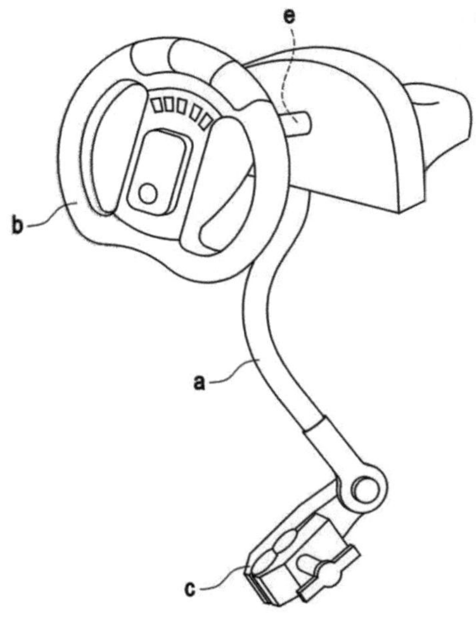

Fig. 3

Fig. 7.19 Flexible mount for a children's toy Fig. 3 (DE 20 2007 001 551 U1)—corrected

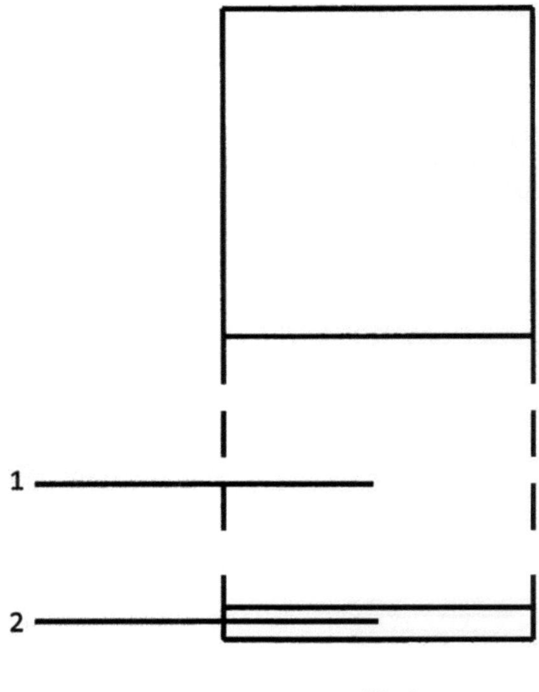

Fig.1

Fig. 7.20 Foldable mobile phone case on textiles Fig. 1 (DE 20 2017 001 820 U1)

Fig.2

Fig. 7.21 Foldable mobile phone case on textiles Fig. 2 (DE 20 2017 001 820 U1)

Table 7.3 Foldable mobile phone case on textiles—list of reference numerals (DE 20 2017 001 820 U1)

Reference list

Fig. 1

1	Inner fabric
2	Closure device

Fig. 2

1	Closure device for opening and closing
2	Closure device for opening and closing to store a cell phone
3	Outer fabric
4	Attachment of the foldable cell phone case

7.3.9 Example: Luminous Band for Baby and Children's Drinking Bottles

Utility model DE 20 2021 000 262 U1 (Title: Luminous band for baby and children's drinking bottles) addresses the technical problem of quickly finding a baby bottle in the dark, for example at night, in order to give a bottle to a crying baby (see Fig. 7.22, DPMA, https://depatisnet.dpma.de/DepatisNet/depatisnet?action=pdf&docid=DE202021000262U1&xxxfull=1, accessed on 23 April 2026).

Arrows may be used in drawings; however, the arrows should start at the reference numeral and end at the element to be identified, or at least point toward the element to be identified. It is permissible to use individual words in a drawing. However, it is more advantageous to use reference numerals, as these can be used in the description and claims to establish a direct reference.

A better representation of the invention can be seen in Fig. 7.23.

Fig. 7.22 Luminous band for baby and children's drinking bottles (DE 20 2021 000 262 U1)

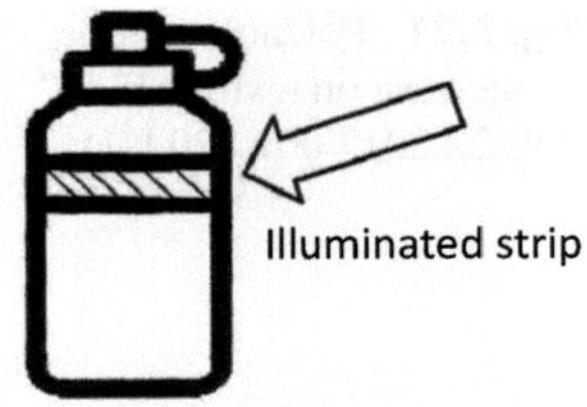

Fig. 7.23 Luminous band for baby and children's drinking bottles (DE 20 2021 000 262 U1)—corrected

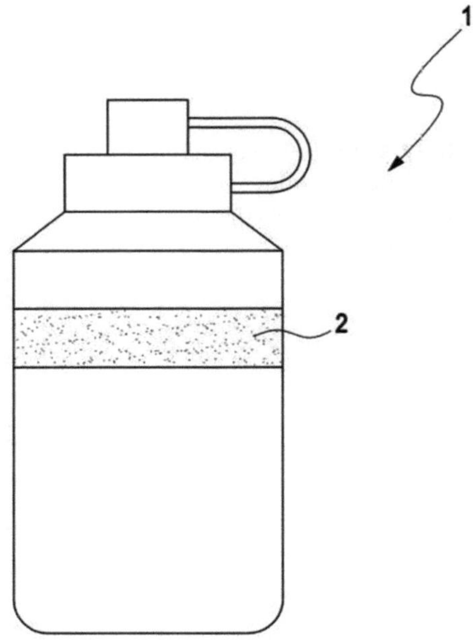

Fig. 1

Description of the Drawings

8

A description of the invention is to be provided based on the drawings. The section "Description of the Drawings" consists of two parts, namely a "Brief Description of the Drawings" and a "Detailed Description of Exemplary Embodiments". In the "Brief Description of the Drawings," each drawing is explained in a single sentence, stating what is depicted in the respective drawing.[1] Additionally, it may be described in which form the elements of the respective drawing are depicted, for example, in a perspective view, a top view, a side view, or an exploded view. After the brief description, a detailed description of specific figures of the invention follows in the section "Detailed Description of Exemplary Embodiments." Here, it is necessary to comprehensively and thoroughly describe what is shown in each drawing and how the individual elements interact to achieve the inventive effect.

8.1 Brief Description of the Drawings

Each figure of the invention is described in a single sentence. This section is introduced with:

> "Exemplary embodiments of the invention are shown in the drawings and are described below. The drawings show:"

or

[1] The drawings are labeled as Fig. 1, Fig. 2, etc.

T. H. Meitinger, *Patent Made Simple: How to File Without a Lawyer*,
https://doi.org/10.1007/978-3-662-73096-6_8

"Further details and advantages of the invention become apparent from the exemplary embodiments shown in the drawings. The drawings show:"

After this, each drawing is briefly explained, in a single sentence. An example follows:

Figure 1 an exploded view of a first embodiment of a harvesting machine according to the invention,
Figure 2 a perspective view of the same harvesting machine as in Fig. 1,
Figure 3 a top view of a second embodiment of the harvesting machine, and
Figure 4 a perspective view of the harvesting machine according to Fig. 3.

The following are two concrete examples from practice:

"Fig. 1 a burnishing tool according to an example of the invention and
Fig. 2 a view of a partial section through a burnishing roller for the burnishing tool from Fig. 1."[2]

and:

"Fig. 1 a drilling machine with a suction device for extracting material removed from a workpiece according to a first exemplary embodiment;
Fig. 2 a side detail view of a drill with a suction sleeve of the suction device from Fig. 1 in the area of a drill tip;
Fig. 3 an oblique view of the drill with suction sleeve from Fig. 2, and
Fig. 4 a front view of the drill with suction sleeve from Figs. 2 and 3."[3]

8.2 Detailed Description of Exemplary Embodiments

After the brief description, a detailed description of exemplary embodiments follows. The detailed description of exemplary embodiments represents an important disclosure of the invention and is usually the place where the invention is explained concretely and comprehensively. It describes which features the invention possesses and how these features interact so that the inventive effects are achieved and the object of the invention is fulfilled. The drawings are provided with reference numerals, and these reference numerals are used in the text of the detailed description to establish a clear connection between the description and the drawings.

[2] DE 10 2019 111 784 A1, DPMA, https://depatisnet.dpma.de/DepatisNet/depatisnet?action=pdf&docid=DE102019111784A1&xxxfull=1, accessed on 23 April 2026.

[3] DE 10 2014 106 968 A1, DPMA, https://depatisnet.dpma.de/DepatisNet/depatisnet?action=pdf&docid=DE102014106968A1&xxxfull=1, accessed on 23 April 2026.

An example from practice:

> *"Fig. 1 shows a perspective view of a burnishing tool 1 for machining a workpiece surface, in particular for roughening or dulling a roughened workpiece surface. The burnishing tool 1 has a base body 12 with a tool section 121 and a mounting section 122 for arrangement in a rotating drive tool (not shown). Eight roller holders 2, arranged radially adjustable, are shown on the tool section 121 of the base body. Each roller holder 2 comprises a mount for burnishing rollers 3, each of which rotates about its own axis of rotation, which is parallel to the base body axis A-A. A burnishing roller 3 is rotatably mounted on each roller holder 2. In the example shown, each or every second burnishing roller 3 has an outer surface with a rough working section (see Fig. 4) with a preset roughness depth, to enable a rolling machining of a workpiece with high machining quality and speed."*[4]

At the end of the detailed description, the reference numerals can be listed. An example follows:

List of Reference Numerals

1. Motor
2. Steering
3. Battery
4. Steering wheel

A list of reference numerals is not mandatory. However, a list of reference numerals is very helpful for better understanding the technical teaching of the application.

In the detailed description of exemplary embodiments, specific embodiments of the invention are explained. It is therefore not absolutely necessary for the subject matter of the main claim to be described. Rather, special embodiments can be presented and explained. The detailed description of exemplary embodiments can be used during the grant procedure to distinguish the invention from the prior art. To achieve this, it should be described which individual features are actually closely linked and can only be used together. Furthermore, it should be explained which features are optional and therefore do not necessarily have to be included in a main claim. This makes it possible to add the optional features to the main claim depending on the identified prior art. Otherwise, all features of the relevant disclosure passage must be included in the main claim, which may result in a very narrow scope of protection. It is therefore advisable to also explain features that can be used alternatively or optionally accordingly.

It is essential to ensure that contradictory statements in the description are avoided. Contradictions between the description and the claims must also be excluded. Otherwise, if the contradictory part of the description is used to interpret a claim, this claim may not be interpretable and therefore unclear.

---

[4] DE 10 2019 111 784 A1, DPMA, https://depatisnet.dpma.de/DepatisNet/depatisnet?action=pdf&docid=DE102019111784A1&xxxfull=1, accessed on 23 April 2026.

The publications of the patent office (published application and patent specification) have consecutively numbered paragraphs. In this case, a sequential number in square brackets is placed at the beginning of each paragraph. This paragraph numbering is carried out by the respective patent office. It is neither required nor desired for the applicant to provide such paragraph numbering.

8.3 Special Formulations

Special formulations are discussed here, which are primarily used in the description of the drawings. Certain aspects must be considered when using these terms. The statements and assessments regarding these terms generally apply to all parts of a patent or utility model application.

8.3.1 "Essential to the Invention"

"Essential to the invention" means that the features so designated are absolutely part of the invention. Labeling a feature as "essential to the invention" results in this feature having to be included in the main claim. By designating a feature as "essential to the invention," the applicant thus binds themselves. This can be disadvantageous in a grant procedure. It may occur that this feature would not need to be included in the main claim in order to sufficiently distinguish from the prior art. In such a case, however, a broader scope of protection cannot be claimed, since the essential feature must be included. The applicant thus imposes an unnecessary limitation on the possible scope of protection from the outset. The designation "essential to the invention" should therefore preferably not be used.

The term "essential to the invention" should also not be used because, during the grant procedure, the direction in which an independent claim is oriented may change. Such a change in direction would theoretically also not be possible if features that are key to this orientation are described as essential to the invention in the description.

8.3.2 Standards and Norms

Applications may contain standards or norms, or the invention may be explained with reference to a norm or standard. Referring to a norm can be problematic, as norms may change over time. If, in addition, no date is given to clarify which version of the norm is meant, the description of the invention may be unclear. It is advisable to describe the invention without references to additional literature, especially normative literature.

8.4 Practical Examples with Recommendations

Below, a portion of the description of the drawings from actual applications is reproduced (in italics). Distinctive passages are shown in bold and underlined. These passages are identified as points ([Point 1], [Point 2], etc.). Comments on these points follow.

8.4.1 Example: Drinking Aid for Mobile and Stationary Use

Utility model DE 20208487 U1 (Title: Drinking aid for mobile and stationary use) describes the technical objective of providing a drinking aid for disabled and elderly persons in such a way that allows them to consume beverages while seated in a normal position. This eliminates the physically strenuous need to bend down to a drinking vessel. An orthopedically disadvantageous posture due to incorrect seating while drinking is thus prevented.

The utility model document contains a drawing (see Fig. 8.1, DPMA, https://depatisnet.dpma.de/DepatisNet/depatisnet?action=pdf&docid=DE000020208487U1&xxxfull=1, accessed on 23 April 2026).

The utility model document provides the following description of the drawings (DE 20208487 U1, DPMA, https://depatisnet.dpma.de/DepatisNet/depatisnet?action=pdf&docid=DE000020208487U1&xxxfull=1, accessed on 23 April 2026):

"Below, an exemplary embodiment of the present invention is ***<u>shown in principle</u>*** [Point 1] . *It shows:"*

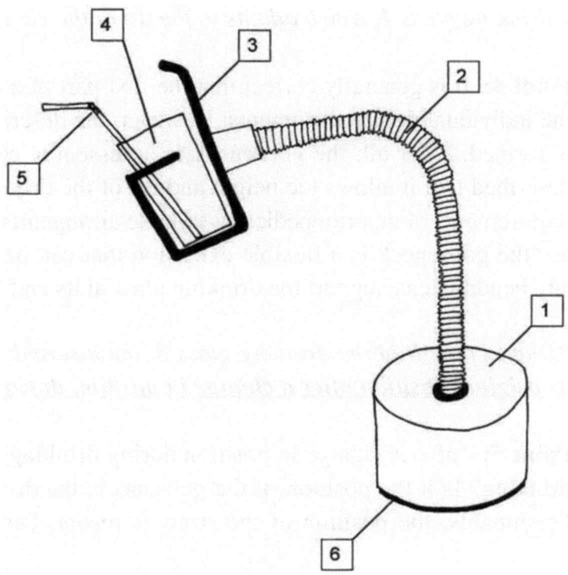

Fig. 8.1 Drinking aid for mobile and stationary use (DE 20208487 U1)

Point 1: "in principle": This term is intended to convey that the exemplary embodiment is depicted and described only in general terms and not in detail. However, the section "Detailed Description of Exemplary Embodiments" is specifically meant to provide a detailed description of concrete embodiments. The abstracted features of the invention are described in the introductory part of the description and in the claims, especially in the main claim.

"Fig. 1 a ***schematic*** [Point 2] *representation of the drinking aid according to the invention with inserted drinking glass and straw."*

Point 2: "schematic": It can be useful to depict only the essential elements in a drawing. Not all parts need to be shown if these additional parts are unrelated to the invention. Alternatively, instead of labeling it "schematic," the type of representation could be described. For example, it could be explained that a perspective view, a side view, a top view, or an exploded view is shown.

"Fig. 1 shows a schematic representation of the drinking aid ***as described below****"* [Point 3].

Point 3: "as described below": redundant phrase, since it is obvious that the drinking aid will be described.

"The movable gooseneck 2 is attached to the base 1. On the other side of the gooseneck 2, the vessel holder 3 is attached. This arrangement allows the height and tilt of the vessel holder 3 to be adjusted as desired. Inserted into the vessel holder 3, as an example, is a drinking glass 4, which adapts to the tilt of the vessel holder 3." [Point 4]

Point 4: It is generally correct that the first part of a drawing description should describe the individual depicted elements. However, the description here omits how the gooseneck is formed. After all, the gooseneck is an essential element of the invention. It could be described that it allows the height and tilt of the drinking glass to be adjusted to meet the requirements of an orthopedically suitable arrangement. For example, an explanation such as "the gooseneck is a flexible extension that can be bent almost arbitrarily and, despite any bending, can support the drinking glass at its end" would be a meaningful addition.

"Due to the tilt of the drinking glass 3, the inserted straw 5 (angled) is also ***returned to its original position after a change in position during drinking****."* [Point 5]

Point 5: "after a change in position during drinking": Which position is changed during drinking? Is it the position of the gooseneck, the drinking glass, or the person drinking? Presumably, the position of the straw is meant. Perhaps the intention is that the straw slides back into the glass after the person has drunk, and for this reason, the tilt enabled by the gooseneck is advantageous. However, this should then be explained accordingly.

"The bottom of the base 1 is ***here*** [Point 6] *provided with a* ***bottom-side*** [Point 7], *non-slip covering 6."*

Point 6: "here": redundant phrase. Meaningless formulations should be omitted in order to produce a patent or utility model document that can be read by a judge, allowing them to quickly and concisely understand the invention. Any inconsistencies or misleading passages are to the detriment of the applicant, as it is their responsibility to describe their invention clearly and understandably.

Point 7: "bottom-side": What is meant by a bottom-side covering? It would be better to state that covering 6 is arranged on the underside of the base. This should be formulated accordingly. Care should be taken to avoid producing unclear passages.

"The illustration also shows that in use ***the gooseneck 2 is also suitable for carrying the drinking aid according to the invention."*** [Point 4]

Point 4: "the gooseneck 2 is also suitable for carrying the drinking aid according to the invention": Contradictory, since the gooseneck is itself a part of the drinking aid. The gooseneck therefore cannot carry the drinking aid, as it is itself a component of it.

8.4.2 Example: Pneumatic Drinking Aid

The utility model DE 20107400 U1 (Title: Pneumatic Drinking Aid) describes a device in which drinking a liquid is facilitated by generating an overpressure in a container holding the liquid (see Fig. 8.2 and Table 8.1, https://depatisnet.dpma.de/DepatisNet/depatisnet?action=pdf&docid=DE000020107400U1&xxxfull=1, accessed on 23 April 2026).

The detailed description of exemplary embodiments of the utility model is as follows (DE 20107400 U1, DPMA, https://depatisnet.dpma.de/DepatisNet/depatisnet?action=pdf&docid=DE000020107400U1&xxxfull=1, accessed on 23 April 2026):

"Technical design of the device [Point 1]*:* ***An air overpressure is generated in a compressed air container (01) with the aid of a pump, compressor, compressed air bottle, etc.*** [Point 2]. *This air overpressure is directed into a vessel (06) containing the liquid to be dispensed (e.g., beer, cola, water, mixed drinks). The vessel is hermetically sealed with a lid (07). Pipes (09) are embedded in this lid, extending almost to the bottom of the vessel (06). The liquid exits the vessel (06) through these pipes (09).* ***There is also an opening in the lid (07) through which the compressed air can flow into the***

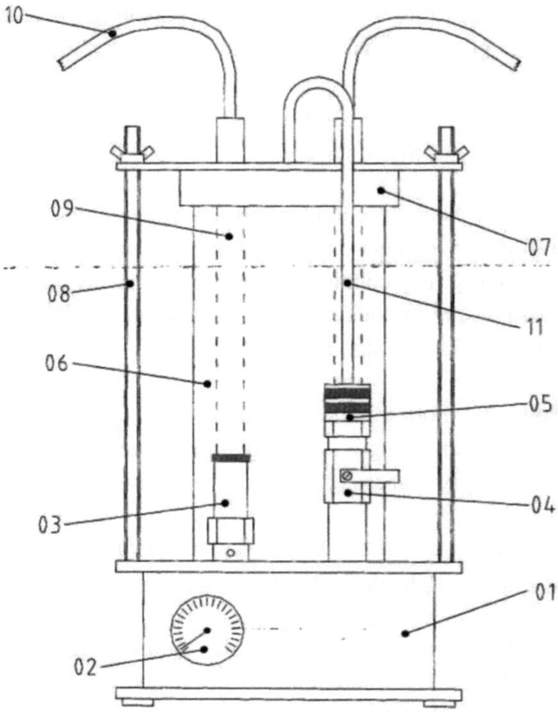

Fig. 8.2 Pneumatic drinking aid (DE 20107400 U1)

Table 8.1 Pneumatic drinking aid – list of reference numerals (DE 20107400 U1)

Components:

01	Compressed air tank
02	Pressure gauge
03	Pressure relief valve
04	Valve
05	Compressed air coupling
06	Vessel
07	Cover
08	Tension rod
09	Pipe
10	Drinking tube
11	Compressed air hose

vessel (06) [Point 3]. *By opening a valve (04) between the compressed air container (01) and the lid (07), the compressed air can suddenly leave the compressed air container (01) and, via the pressure hose (11) and lid (07), enter the vessel (06), where it acts on the liquid to be dispensed. The overpressure in the vessel (06) forces the liquid out through the pipes **(09)*** [Point 4] *and the attached drinking tubes (10). To monitor the* *pressure, a manometer (02) and a pressure relief valve (03) are attached to the compressed air container (01).* ***For quick disconnection of the lid (07) from the compressed air container (01), a compressed air coupling (05) is installed between the valve (04) and the compressed air hose (11)*** [Point 5]."

Point 1: "Technical design of the device": Unnecessary phrase. A patent or a legally valid utility model can only be granted for a technical invention.

Point 2: "An air overpressure is generated in a compressed air container (01) with the aid of a pump, compressor, compressed air bottle, etc.": Reference numeral 01 points to the lower section of the device. Is this section supposed to be a compressed air container, i.e., a container capable of holding compressed air? If so, why is a "compressed air bottle" mentioned as being "in the compressed air container," since the lower section is already a vessel that can hold compressed air?

Point 3: "There is also an opening in the lid (07) through which the compressed air can flow into the vessel (06)": How does this opening interact with the compressed air container 01, in which the compressed air is generated, and why is there an opening in the lid to introduce compressed air into vessel 06? Or is the opening in lid 07 used to introduce compressed air into vessel 06 from an external source, such as a compressed air bottle, instead of from compressed air container 01? The description of the invention must be clear!

Point 4: "(09)": Reference numerals should be used without a leading zero. Furthermore, reference numerals should not be placed in parentheses in the description.

Point 5: "For quick disconnection of the lid (07) from the compressed air container (01), a compressed air coupling (05) is installed between the valve (04) and the compressed air hose (11)": Why or when is quick disconnection needed? How do the elements interact? Is valve 04 closed first or not? The individual elements and their functions and interactions are discussed far too little. For example, there is also no description of which components of the invention are located where, i.e., inside or outside vessel 06 or compressed air container 01.

9 Claims

The drafting of claims requires great care. The claims determine what is protected.[1] It is important to claim the broadest possible scope of protection. Otherwise, there is a risk that workarounds will be possible. In such a case, the invention has been disclosed to the public through the application, without being able to prevent commercial imitation of the invention.

A claim is only allowable if it is novel and inventive. Therefore, when drafting claims, one should be familiar with the relevant prior art documents. A claim is novel if its features are not disclosed in a single location in a prior art document.[2] The claim is based on an inventive step if its features are not obvious for the person skilled in the art from the prior art documents.[3]

It is therefore advisable to conduct your own search for prior art documents in order to be able to draft meaningful claims. This enables a good start in the grant procedure before the patent office.[4]

Nevertheless, it will usually be necessary to adapt the claim wording during the grant procedure. It should be kept in mind that the examination procedure serves to find the appropriate, i.e., the broadest possible, claim wording. "Broadest possible" has two dimensions: the broadest and the allowable scope of protection. Once an allowable scope of protection has been found, the main claim should fully cover this scope. Alternative

---

[1] The first claim is called the main claim. The main claim, together with the independent claims, determines the scope of protection. Each of these claims defines a separate scope of protection. The prerequisite for this is, in particular, that these claims are novel and inventive.

[2] Section 3(1) sentence 1 of the Patent Act or Article 54(1) EPC.

[3] Section 4 sentence 1 of the Patent Act or Article 56 sentence 1 EPC.

[4] See Chap. 12 "Search for Prior Art".

T. H. Meitinger, *Patent Made Simple: How to File Without a Lawyer*,
https://doi.org/10.1007/978-3-662-73096-6_9

main claims would result from the further inclusion of additional features. However, this would reduce the scope of protection and allow for workarounds by imitators. It is therefore important to claim the broadest possible scope of protection through the main claim and the independent claims. This search for the broadest possible scope of protection takes place through correspondence with the examiner or the examining division of the patent office during the grant procedure.[5]

There is no examination procedure in utility model law. It is only possible to request that the patent office determines the prior art to be considered in assessing legal validity.[6] As a result, utility models can also be registered for items that are clearly not novel and inventive, for example, a car with four wheels. Although a utility model has been registered by the patent office, it may nevertheless not be legally valid.

The scope of protection of a patent is defined by the set of claims. The set of claims includes a main claim, possibly independent claims, and dependent claims. The dependent claims always refer back to the main claim or an independent claim and, by virtue of this reference, include all the features of the main claim or the independent claim. In addition, the dependent claims contain further features that are not included in the main claim or the independent claim. Since the features of the main claim or the independent claim are included in the dependent claims, it follows that the scope of protection of a dependent claim cannot be broader than that of the main claim or the independent claim. Due to the additional features of the dependent claim compared to the main claim or the independent claim, the scope of protection of the dependent claim is generally smaller than that of the main claim or the independent claim. The scope of protection of a dependent claim can therefore at most be equal to that of the referenced main claim or independent claim. The prerequisite for equality of scope is that the additional features of the dependent claim do not have a technical character. The scopes of protection of the dependent claims are therefore necessarily never broader than those of the main claim or the independent claims. When determining the scope of protection of a patent or a utility model, it is therefore sufficient to analyze the main claim and the independent claims.

Note: Only the independent claims (main claim and independent claims) determine the scope of protection. Each independent claim defines a separate scope of protection. In the case of a patent, these are called patent claims, and in the case of a utility model, protection claims. The entirety of the claims is referred to as the set of claims.

If possible, several independent claims should be included in a set of claims. For example, a set of claims may contain a device claim and a method claim describing how the device is manufactured or used.

Only those features that are essential should be included in the main claim and an independent claim. Only the features without which the invention cannot be carried out

[5] Section 45(2) of the Patent Act.

[6] Section 7(1) of the Utility Model Act.

should be described. Each additional feature would reduce the scope of protection, since only those products or processes that implement all the features of a claim constitute patent infringement.

By including features in a claim, the threshold for infringement is raised, since at least potentially each feature reduces the scope of protection. Therefore, attention must be paid to every single feature and every single word to ensure that there is not already an unnecessary limitation of the scope of protection. A typical example is the attachment of a first element, for example, a picture, to a second element, for example, a wall. The claim may specify that the connection is made with a nail. However, it may not matter what type of connection is used, but only that a connection exists at all. In this case, it may be advantageous to specify that a fastening means holds the two elements together. In a dependent claim, it can be specified that the fastening means is a nail, a hook, a screw connection, or an adhesive. It can also be specified whether the connection is detachable or non-detachable.

Every single word of a claim, especially of the main claim and the independent claims, must be carefully examined for its necessity or required level of abstraction. The limit is the prior art. A claim in an application must be drafted so that, in light of the prior art, it is just novel and inventive.

Tip: The fewer features a claim has, the broader its scope of protection tends to be. Each feature further specifies the subject matter of the invention and thus potentially reduces the scope of protection. The subject matter of the claim therefore becomes smaller as the number of features increases. Therefore, especially in the independent claims, only those features should be included that are necessary for the invention to be carried out. An independent claim therefore includes only those features that are required for the invention to be (just) realizable.

A claim must be drafted so that the claimed invention can be carried out. However, no unnecessary feature should be included in the independent claims. Otherwise, the invention originally created by the inventor can be implemented by a competitor, in particular without the unnecessary feature. However, it may turn out during the grant procedure that the original invention is not allowable. In this case, further features must be included. It should be borne in mind that the grant procedure is precisely intended to achieve an optimal scope of protection. This means the broadest possible scope of protection in light of the prior art.[7]

---

[7] For a patent application, prior art refers to all knowledge that was made available to the public before the filing date by written or oral description, by use, or in any other way (Section 3(1) sentence 2 of the Patent Act). For a utility model application, a narrower definition of prior art applies. Oral descriptions are not taken into account, and uses are only relevant if they occurred within the country (Section 3(1) sentence 2 of the Utility Model Act). In addition, a general novelty grace period applies to utility models (Section 3(1) sentence 3 of the Utility Model Act).

Tip: In the dependent claims, additional features that are not directly essential to the invention can be included.

Dependent claims serve as possible fallback positions that can be used to render an independent claim legally valid by incorporating the features of a dependent claim. This allows the previous independent claims to be reformulated and distinguished from the prior art. To avoid unnecessary costs, multiple features can be combined in a single claim using "and/or". In this way, the number of claims can be limited to 10 or 15.[8]

The entirety of the claims of a patent application, a patent, or a utility model application is referred to as the set of claims. Two types of claims can be distinguished: independent claims and dependent claims. The dependent claims refer back to other claims (reference: "… according to one of the preceding claims, wherein …"). An independent claim does not contain such a reference. The independent claims include the main claim (claim 1) and any parallel claims. These independent claims define the scope of protection of the patent or utility model.

An independent claim begins with a designation, for example, device for …, method for …, harvesting machine for …, construction machine for …, etc., and possibly further characteristics of the device, method, harvesting machine, construction machine, etc. Typically, this is followed by a list of features required to implement the invention. The list can be introduced by the terms "comprising," "having," or "with":

1. Device/method for …, comprising:
   - Feature 1 for …,
   - Feature 2 for … and
   - Feature 3 for …,

   characterized in that/wherein
   the device/method comprises/has:
   - Feature 4 for …

Features 1, 2, and 3 are prior art and not new. After a "characterized in that" or "characterized by," one or more features follow that are new and, in combination with features 1, 2, and 3, produce the inventive effect and thus lead to inventive activity.

9.1 Recommended Examples from Practice

In document DE 10 2019 111 784 A1, the first claim is formulated as follows (see Figs. 9.1 and 9.2, DPMA, https://depatisnet.dpma.de/DepatisNet/depatisnet?action=pdf&docid=DE102019111784A1&xxxfull=1, accessed on 23 April 2026):

[8] In German proceedings, a claim fee is due from the 11th claim onward, and in European proceedings, from the 16th claim onward.

Fig. 9.1 Burnishing tool Fig. 1 (DE 10 2019 111 784 A1)

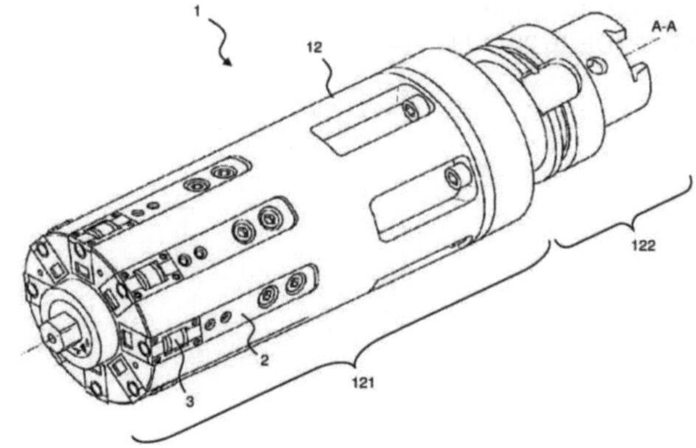

"1. Burnishing tool (1) for machining a workpiece surface, comprising a base body (12) with at least one roller holder (2) arranged to be radially adjustable, wherein a burnishing roller (3) is rotatably mounted on each roller holder (2), and wherein at least one burnishing roller (3) has an outer surface with a rough working section (4) with a predetermined surface roughness depth."

The first claim, the main claim, of a set of claims is typically the most important claim. The main claim defines the scope of protection. If additional independent claims are included in the set, further scopes of protection arise.

This claim 1 begins with the designation of the subject matter to be protected, namely a burnishing tool. Alternatively, one could begin with "device" or "method."

It also describes the intended purpose of the burnishing tool, namely "for machining a workpiece surface." This statement of purpose is not limiting, but it helps to clarify what the burnishing tool is intended for. Purpose statements (for …) can also be advantageously used for individual features of a claim. This can help to clarify the interaction of the features.

Document DE 10 2019 111 784 A1 (DPMA, https://depatisnet.dpma.de/DepatisNet/depatisnet?action=pdf&docid=DE102019111784A1&xxxfull=1, accessed on 23 April 2026) also includes the following independent method claim:

"9. Method for producing a burnishing roller (3) comprising the following steps:

- *Providing a burnishing roller (3) with a working section (4);*
- *Roughening a surface of the working section (4) by means of*
  a) *laser and/or*
  b) *grinding and/or*
  c) *coating."*

It is not uncommon for a set of claims to include, as the main claim, a product claim and, as an independent claim, a method claim. The method claim usually describes how the subject matter of the main claim is manufactured or how the subject matter is used.

Fig. 9.2 Burnishing tool Fig. 2 (DE 10 2019 111 784 A1)

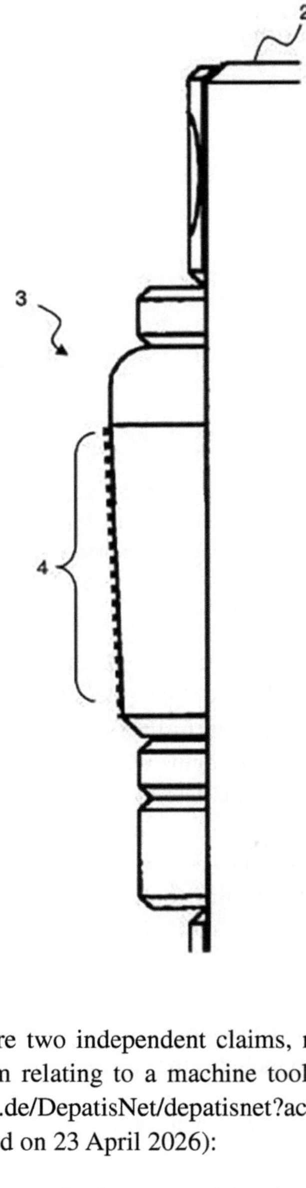

Fig. 2

In document DE 10 2014 106 968 A1, there are two independent claims, namely a device claim for an extraction device, and a claim relating to a machine tool with an extraction device (DPMA, https://depatisnet.dpma.de/DepatisNet/depatisnet?action=pdf&docid=DE102014106968A1&xxxfull=1, accessed on 23 April 2026):

> *"1. Extraction device (13) of a machine tool (10) for extracting loose material such as chips and the like, which can be removed from a workpiece (12) by a tool (22) of the machine tool (10) rotating about a tool axis (18), comprising at least one extraction channel (58) which opens to at least one working section (24) of the tool (22), which leads into a collection area (42) remote from the at least one working section (24), and which can carry a transport fluid flowing away from the at least one working section (24), characterized in that the at least one extraction channel (58) is arranged at least in sections eccentrically with respect to the tool axis (18) in and/or on the tool (22)."*

> *"12. Machine tool (10) with at least one extraction device (13), for extracting loose material such as chips and the like, which can be removed from a workpiece (12) by at least one tool (22) of the machine tool (10) rotating about a tool axis (18), comprising at least one extraction channel (58) which opens to at least one working section (24) of the tool (22), which leads into a collection area (42) remote from the at least one working section (24), and which can carry a transport fluid flowing away from the at least one working section (24), characterized in that the at least one extraction channel (58) is arranged at least in sections eccentrically with respect to the tool axis (18) in and/or on the tool (22)."*[9]

An "extraction device of a machine tool" is claimed, as well as the "machine tool" itself. It is not unusual for a set of claims to include several product claims that essentially claim increasingly larger units. For example, an engine can be claimed, and additionally a vehicle that comprises this engine. For this, a claim can be formulated: "6. Vehicle comprising an engine according to any one of claims 1 to 5 …". This approach advantageously defines not only the infringing object of the engine, but also that of a vehicle. By this means, an infringement of the intellectual property right can potentially be demonstrated more clearly.

9.2 Claim Categories

The subject matter of a claim can be assigned to different claim categories (see Fig. 9.3). According to the EPC (European Patent Convention), there are four claim categories: product claims, process claims, apparatus claims, and use claims.[10] It is also conceivable to consider manufacturing claims. Product claims and apparatus claims can be grouped together as substantive claims or object claims. Process claims, use claims, and manufacturing claims can be referred to as process claims, since the use of an object or the manufacture of an object is carried out by a process.

The categorization of claims has direct practical relevance, since an independent claim from each category can be included in a set of claims. Advantageously, each independent claim defines its own scope of protection,[11] so that a maximum of four independent scopes of protection can be defined with each patent.

It is permissible in a set of claims to direct an independent claim to a product and to describe an independent claim that specifies a specially adapted process for manufacturing the subject matter of the product claim. Additionally, an independent claim for the use of the product may be included in the set of claims (use claim).

[9] DE 10 2014 106 968 A1, DPMA, https://depatisnet.dpma.de/DepatisNet/depatisnet?action=pdf&docid=DE102014106968A1&xxxfull=1, accessed on 23 April 2026.

[10] Rule 43(2) EPC.

[11] Section 14 sentence 1 Patent Act or Article 69(1) sentence 1 EPC.

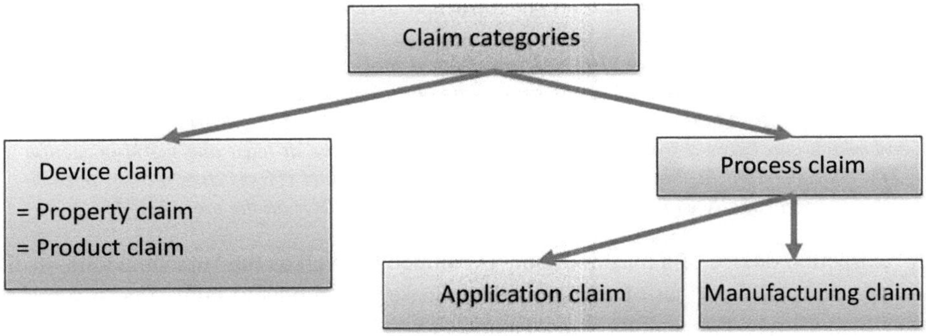

Fig. 9.3 Claim categories

It is also permissible in a patent application to claim a process in an independent claim and to direct another independent claim to a specific apparatus or special means for carrying out this process.

Another variant for including several independent claims in a set of claims would be to formulate an independent claim for an inventive product and to describe a specially adapted process for the application of the product. In addition, an apparatus or means developed for carrying out the specially adapted process could also be claimed.

The use of multiple independent claims is only permissible if the subjects of the independent claims fulfill the requirement of unity of invention. Unity exists if only one invention is claimed by the claims, or if different inventions are claimed that are, however, linked by a single inventive concept.[12] An example is an inventive product for which a special apparatus for its manufacture or application has been developed. Another example is an inventive process for which an apparatus for its application has been specially developed. Independent claims can, in particular, be linked by common technical features that contribute to technical progress.[13] It is not required that these common technical features are new. However, their specific application must not have been previously known.

The variants involving manufacturing or application processes exist exclusively for patents and not for utility models, since a process cannot be claimed in a utility model. In a utility model, the manufacture or use of an object cannot be claimed.[14] Therefore, in a utility model claim, only a claim to an inventive product can be directed, and additionally to an apparatus that has been specially developed to apply or manufacture this product.

[12] Section 34(5) Patent Act or Article 82 EPC.

[13] Rule 44(1) sentence 2 EPC.

[14] Section 2 no. 3 Utility Model Act.

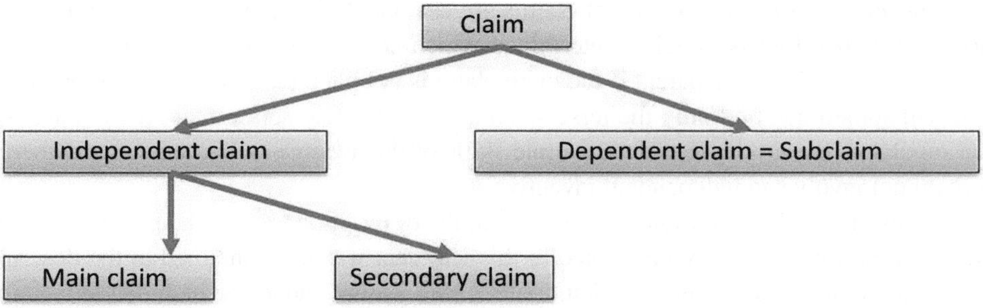

Fig. 9.4 Independent and dependent claims

9.3 Independent and Dependent Claims

There are independent, dependent, and co-ordinate claims. In addition, there are subclaims and a main claim (see Fig. 9.4). The main claim is the first claim listed in the set of claims. The main claim contains the essential features of the invention.[15] The independent claims are the co-ordinate claims and the main claim. The dependent claims are the subclaims.[16]

Dependent claims are claims that contain all the features of another claim.[17] Co-ordinate claims are alternatively referred to as parallel claims.[18] Each parallel claim defines its own scope of protection, in addition to that of the main claim. For example, parallel claims may be process claims if the main claim is, for example, an apparatus claim. A parallel claim may refer to a preceding claim. For example, a process claim may refer back to preceding apparatus claims by the wording "Process for using an apparatus according to one of the preceding claims…". In this case, all the subjects of the apparatus claims are incorporated into the process claim and can be used without separate explanation.

Alternatively, a co-ordinate apparatus claim may refer back to preceding process claims by the following wording: "Apparatus for carrying out a process according to one of the preceding claims…".

The independent claims define the scope of protection. The dependent claims serve as fallback positions in a patent grant procedure if the independent claims are not allowable. Therefore, promising candidates for patenting should be included in the dependent claims, since in an official communication during an examination procedure before

[15] Section 9(4) Patent Ordinance.
[16] Section 9(6) sentence 1 Patent Ordinance.
[17] Rule 43(4) sentence 1 EPC.
[18] Section 9(5) sentence 1 Patent Ordinance.

a patent office, all claims of a set of claims are evaluated. By analyzing the communication, it can therefore be quickly determined which combination of a subclaim with the main claim may lead to a patent if the main claim is not patentable. However, in practice, one will sometimes find only the terse statement in an office action that "the features of the subclaims merely represent the routine skills of the average skilled person and therefore do not involve an independent inventive step."

It is also possible to "elevate" features from the description into an independent claim. However, nothing new may be created in this way that was not contained in the description. If a feature is described in a first context, this feature cannot be included in a claim that represents a different second context. Otherwise, the feature would be used in a way that distorts the meaning of the description, and this would constitute an inadmissible extension.[19] For example, if a feature is mentioned at one point, while it is simultaneously stated that this feature is only suitable for an electric vehicle, this feature cannot be used in a claim directed to an internal combustion vehicle. This also applies even if the feature could in fact be used in an internal combustion vehicle, since the inventor did not describe this variant in the originally filed application.

9.4 Dependent Claims

With the features of the dependent claims, the invention is further developed in a technically advantageous manner. Essentially, only features that contribute to inventive activity should be described here. Trivialities that are known to any skilled person do not belong in the dependent claims. Therefore, dependent claims should not contain self-evident matters, but rather technically advantageous further developments of the invention. After all, the features of the dependent claims are intended to serve as fallback positions which, in combination with the main claim—should the features of the main claim alone not suffice—establish novelty and inventive step.

9.5 Structure of a Set of Claims

A set of claims may include a main claim and several independent claims. Dependent claims may refer to each of these claims. The main claim is written first. This is followed by the dependent claims that refer to the main claim. Next come the independent claims, if any. After the independent claims, the dependent claims referring to them are

[19] Section 38 sentence 2 Patent Act (no rights from an inadmissible extension of the application) or Section 21(1) no. 4 (revocation of the patent in case of inadmissible extension) or Article 123(2) EPC (inadmissible extension of the application) and Article 123(3) EPC (inadmissible extension of the scope of protection).

arranged. The complete set of claims, including the main claim, independent claims, and dependent claims, is numbered consecutively with Arabic numerals.[20] The main claim is assigned the first numeral "1".

9.6 Reference of Claims

A dependent claim must refer back to at least one claim. Typically, this can be done using a formulation such as "claim according to one of the preceding claims …".[21]

The following example of a set of claims comprises ten claims with five device claims and five method claims. The main claim is a device claim, to which four dependent claims refer (claims 2 to 5). In addition, the set of claims includes a method claim as an independent claim (claim 6). Four dependent claims refer back to this independent claim (claims 7 to 10).

1. Device for …
2. Device according to claim 1, comprising …
3. Device according to one of claims 1 or 2, comprising …
4. Device according to one of the preceding claims, comprising …
5. Device according to one of the preceding claims, comprising …
6. Method for applying/manufacturing the device according to one of the preceding claims, comprising the steps: …
7. Method according to claim 6, further comprising the steps: …
8. Method according to one of claims 6 or 7, further comprising the steps: …
9. Method according to one of claims 6 to 8, further comprising the steps: …
10. Method according to one of claims 6 to 9, further comprising the steps: …

With regard to formulating the reference, it is possible to simplify matters by always using the phrase "according to one of the preceding claims." In individual cases, however, this may be incorrect if a dependent claim uses a subject matter that, for example, is only introduced in claim 3 and not already in claim 1. Such an error can be corrected at any time during the patent grant procedure or in litigation. However, it is not possible to extend a reference to claims to which the relevant claim did not originally refer.

Tip: A reference should initially be made to all preceding claims. For the dependent claims of the main claim, the formulation "according to one of the preceding claims" is therefore recommended. For dependent claims of an independent claim, the formulation "according to one of claims 3 to 6" is suitable, for example if the independent claim is claim 3 and the formulation is used for dependent claim 7.

---

[20] Section 9, paragraph 7 of the Patent Ordinance.

[21] Section 9, paragraph 6, sentence 2 of the Patent Ordinance.

9.7 Single-Part or Two-Part Claim Format

In the two-part claim format, the use of the terms "characterized by" or "characterized in that" divides the claim into the so-called preamble or generic term and the characterizing portion. The preamble corresponds to the closest prior art, i.e., what was previously known and from which the inventor started when creating the invention. The characterizing portion describes what is new and constitutes the inventive step.

A prerequisite for dividing into preamble and characterizing portion is that the relevant prior art is known. However, this can only be determined with sufficient certainty during the ongoing examination procedure before the patent office. Therefore, it is not advisable for the applicant to divide the claim into preamble and characterizing portion already when drafting the application. Moreover, a corresponding "characterized …" formulation can be added at any time during the grant procedure before the patent office. In any case, this avoids an initial misjudgment of the prior art by the applicant, which could be detrimental. It is quite possible that an examiner initially relies on the applicant's misjudgment of the prior art and thus underestimates the inventive core of the invention. This can at least lead to a prolongation of the examination procedure.

Furthermore, a combination invention may be present, in which all features of the claim are known individually, but the combination of the individually known features constitutes the inventive aspect. In this case as well, a division into preamble and characterizing portion should be avoided.

It is generally up to the applicant to choose whether to use a single-part or two-part claim format.[22] In light of the above considerations, it is recommended to use the single-part claim format.

9.8 Definite and Indefinite Articles

When a term or object is introduced for the first time in a set of claims, the indefinite article should be used. When a term is referred to again in a dependent claim, the definite article should be used, since the term or object has already been introduced. The following is an example in which a bicycle is introduced in the main claim and a sidecar in claim 3.

1. Device for …, comprising <u>a</u> bicycle.
2. Device according to claim 1, wherein <u>the</u> bicycle …
3. Device according to one of claims 1 or 2, comprising <u>the</u> bicycle and <u>a</u> sidecar …
4. Device according to claim 3, wherein <u>the</u> sidecar …

[22] Section 9, paragraph 1, sentence 1 of the Patent Ordinance.

9.9 Claim Fees

If a certain number of claims is exceeded, claim fees become due. For a German patent, a fee of 30 EUR must be paid starting from the 11th claim.[23] For a European application, claim fees are due starting from the 16th claim. For each claim after the 15th, 290 EUR must be paid. From the 51st claim onwards, the claim fee is even 720 EUR per claim.[24]

It therefore makes sense to draft a maximum of 10 or 15 claims. If an invention has more than ten specific features, two or more features can be combined in a dependent claim using an "and/or" connection. Otherwise, these details can be listed in the description, where they can, if necessary, be incorporated into the main claim should a distinction from the prior art require it.

The maximum number of claims that can be filed without an additional fee should be fully utilized. This is because patent office examiners are required to provide an assessment of patentability for all claims during examination. This may quickly reveal which features of a dependent claim could allow for a grant of the patent. Therefore, it makes no sense to include technical trivialities in the dependent claims that, in combination with the features of the main claim, have no chance of being patentable. The selection of features for the dependent claims should be made with care.

9.10 Interpretation of Claims

The meaning of a claim is its technical teaching, which forms the basis both in the grant procedure and in infringement proceedings. Based on the meaning, it can be determined whether a claim is novel and inventive, or whether a product actually infringes the scope of protection of a patent or not. As a rule, a claim is understandable to the skilled person on its own. However, it may happen that a claim is not self-explanatory and must then be interpreted with reference to the description and the drawings.[25] Interpretation can also be carried out with the help of the other claims.

It is advisable to draft the claims in such a way that the description and drawings are not required to understand the claims. Otherwise, an element of uncertainty arises, since the need for interpretation introduces a certain degree of ambiguity. This can result in a patent office examiner attributing a broad meaning to a claim during examination, thereby endangering novelty or inventive step. Conversely, in infringement proceedings, an attempt may be made to assign a narrow scope of protection to a claim, which may

[23] DPMA, https://www.dpma.de/service/gebuehren/patente/index.html, as of March 4, 2021.

[24] EPO, https://www.epo.org/law-practice/legal-texts/official-journal/2020/etc/se3/2020-se3.pdf or https://my.epoline.org/epoline-portal/classic/epoline.Scheduleoffees, as of March 4, 2021.

[25] Section 14 sentence 2 Patent Act or Article 69 paragraph 1 sentence 2 EPC.

lead to a finding of non-infringement. The applicant is therefore well advised to draft claims so that the meaning of a claim is conveyed clearly and unambiguously.

It would be disastrous if a term in a claim is to be understood in one way in the claim and is defined differently in the description. In such a case, an ambiguity arises, which is always interpreted to the detriment of the applicant.

A claim is to be understood from the perspective of a person skilled in the art. The skilled person will assign to the individual terms the meaning that results from the technical teaching described in the claim. If the skilled person derives a coherent teaching from the claim, they will not consult other parts of the application. Otherwise, the skilled person will look for clues for interpretation in the other claims, the description, and the drawings.

Interpretation can be carried out in three steps. First, the meaning of the individual terms of a claim is determined. This can already lead to ambiguities, since a term can have different meanings. For example, "Spannung" can mean electrical voltage or mechanical stress. In a second step, the context of the claim is used to determine whether electrical or mechanical stress is meant. If this is not successful, the other claims, the description, and the drawings are consulted.

The issue of interpretation arises in particular with legal texts. Four fundamental methods of interpretation have developed here:

- **Grammatical, literal or. semantic interpretation:** Literal interpretation interprets a legal provision according to the meaning of the individual words. The interpretation of the legal provision thus corresponds to the concatenation of the respective meanings of the words used.
- **Systematic interpretation:** A systematic interpretation is one that results from the context. The individual words of a legal provision are assigned those meanings that arise from the context.
- **Historical interpretation:** In historical interpretation, the meaning that results from the legislative history is sought.
- **Teleological interpretation:** In teleological interpretation, the meaning and purpose of the legal provision are sought.

A claim can generally be interpreted using a semantic or systematic approach. Here, interpretation must be carried out from the perspective of an average skilled person. Historical or teleological interpretation, which is certainly possible for a law, are excluded for a claim. The legislative history is not taken into account. Likewise, a patent office will not attempt to determine what the applicant might have meant. The applicant should ensure that their description and claims are clearly drafted. An infringement court will likewise not undertake a teleological interpretation of the claims or award damages based on conjecture. The infringement court expects the applicant to clearly and unambiguously define what is claimed as their intellectual property.

It is therefore of great importance to draft claims clearly and precisely. Otherwise, the "patent" or "utility model" project is doomed to fail from the outset, and although the applicant will disclose their know-how through publication of their invention in the application publication, they will not be able to prevent a competitor from imitating their invention. In summary, good patents protect and bad patents harm, with bad patents being particularly characterized by deficient claims.

In claim formulations from practice, a complicated linguistic style can often be observed, which may be intended to impress through its lofty and seemingly legalistic manner. The use of such a writing style can only be discouraged. It regularly leads to overly narrow formulations, enabling workarounds, or to overly abstract passages that do not withstand scrutiny for feasibility. In addition, formulations may arise that are simply not understandable and therefore do not clearly and precisely specify the features of the invention. Such ambiguity is always to the detriment of the applicant, since in the grant procedure a broad interpretation is applied, requiring a claim to be distinguished from many prior art documents. Conversely, in infringement proceedings, a narrow interpretation of unclear terms is applied, which may result in denying infringement by an imitator.

9.11 Non-Technical Features

The Patent Act is only applicable to technical inventions.[26] An invention must therefore possess a technical character. However, a claim can also be patentable if it contains non-technical features.[27] Nevertheless, the subject matter of the claim as a whole must exhibit a technical character.[28]

The concept of a "non-technical" feature is imprecise, as a feature may be considered technical in one context and non-technical in another. For example, painting a device with red paint. The red coating may enable the device to be clearly identified by a camera system. In this case, the red coating constitutes a technical feature. In another case, a red coating on a device may serve only an aesthetic purpose and thus would not be a technical feature.[29]

Such ambiguity regarding technicality occurs relatively often. The technicality of a feature often only emerges from its interaction with the other features of the relevant

[26] Section 1, Paragraph 1 of the Patent Act.

[27] Technical Board of Appeal EPO, T 26/86 'X-ray device', OJ EPO 1988, 19.

[28] Stauder/Luginbühl, European Patent Convention, Commentary, 7th edition, Article 52, margin note 16.

[29] Section 1, Paragraph 3, No. 2 of the Patent Act (for aesthetic creations, design protection can be obtained under the Design Act).

claim. For example, a tire tread may improve braking performance or merely appeal to aesthetic preferences. A vessel may serve purely aesthetic purposes or provide a necessary liquid for the operation of a device. The surface of a house wall may be designed in a particular way purely for visual reasons or may serve the function of better withstanding weather conditions. The question of technicality is particularly relevant in software applications.[30]

Non-technical features are not considered in the interpretation of the claim, but are simply ignored. However, non-technical features carry the risk of being interpreted as technical, even though the inventor never intended this. In such cases, these features can drastically reduce the scope of protection of the relevant claim. The applicant should therefore carefully check their claims to ensure that they actually contain only technical features.

9.12 No Equivalent Infringement

An equivalent infringement of a patent or utility model occurs if a form of infringement does not literally implement the features of an independent claim of the patent or utility model, but does so equivalently. Equivalence requires the presence of three criteria: same effect, equivalence in value, and obviousness to the skilled person. The legal concept of equivalent infringement would result in an extension of the scope of protection beyond literal infringement. Currently, infringement courts consider such an extension to be inadmissible, as it would create legal uncertainty. At present, infringement court recognizes equivalent infringement very reluctant. An argument by the patent owner such as "… the inventor also meant …" will therefore not succeed. It is all the more important to draft claims as abstractly as possible and as specifically as necessary.

The applicant must therefore ensure, in their own interest, that their claim formulations are clear both in terms of wording and in the context in which they are used. A prerequisite for this is a critical distance from one's own formulation. The applicant should bear in mind that a claim is an instruction to a person skilled in the art to implement the invention. The skilled person has only their industry-standard expertise at their disposal. The description of the invention is initially not taken into account. The claims must therefore clearly, unambiguously, and precisely specify how the skilled person is to proceed.[31] On the other hand, under no circumstances should independent claims include

[30] Software as such is not eligible for patent protection (Section 1, Paragraph 3, No. 3 and Paragraph 4 of the Patent Act).

[31] EPO, Guidelines for Examination, Part F, Chapter IV, 4.1—Clarity (retrieved April 14, 2021, https://www.epo.org/law-practice/legal-texts/html/guidelines/d/f_iv_4_1.htm).

features that are not absolutely necessary for the implementation of the invention. Features that are unnecessarily included in the main claim or a parallel claim unnecessarily reduce the scope of protection and provide imitators with opportunities to circumvent the claim. The applicant should therefore exercise the utmost care in drafting the claims.

9.13 Claim Drafting through Abstraction

If a claim describes a specific embodiment, the claim is limited to this specific embodiment. It is irrelevant whether there are similar, equivalent, and well-known variants of the specific embodiment to the skilled person. A typical example is a nail for fastening a device to a wall. If the nail is described in the claim, the scope of protection is limited to the device with a fastening by a nail to the wall. Fastening by a dowel or a hook-and-loop fastener would not constitute patent infringement. It is therefore important to ensure that a claim, especially the independent claims that determine the scope of protection, encompasses all obvious variants through its individual features. In the specific example of the nail, it would be sensible to include the feature "fastening means" or "means for fastening."

A prerequisite for the use of a means-plus-function feature is that the feature is understandable and executable for the skilled person. In this case, the feature encompasses all variants that perform the described function. A "means for fastening" could, for example, be an adhesive strip, a screw, a dowel, a hook-and-loop fastener, or a nail.

Drafting claims with means-plus-function features has the advantage that the physical, structural features do not need to be described. Instead, only the effects of the individual elements of a device according to the invention are defined. This results in an abstraction of the claim wording that encompasses all items with the same effect. It is no longer necessary to mention items with the same effect separately. However, there is a risk that means-plus-function claims may be regarded by a patent office or an infringement court as being formulated in terms of the problem to be solved. In order to avoid the objection of a problem-based formulation, three requirements must be met:

Sufficient Disclosure Concrete embodiments of the means-plus-function features must be explained in the description. This ensures that means-plus-function features are executable for the skilled person. For example, in the introductory part of the description, it could be stated directly after the relevant claim that a means for fastening or a fastening means is a nail, a hook, a pin, or an adhesive compound.[32]

[32] EPO, Guidelines for Examination, Part F, Chapter IV, 6.5—Functional Definition, (retrieved on March 17, 2021, https://www.epo.org/law-practice/legal-texts/html/guidelines/d/f_iv_6_5.htm).

No Concrete Description Possible Furthermore, it must be impossible or at least difficult to describe the specific embodiments that fall under a means-plus-function feature themselves. In this case, describing the feature by its effect is therefore the more elegant way of drafting.

Description of the Solution The means-plus-function formulation actually represents a description of the solution means and not merely an explanation of the problem.

When drafting claims, it must also be taken into account that the respective independent claim defines the scope of protection and that this scope cannot be extended by dependent claims referring back to this independent claim. It is therefore not possible to describe a nail in the main claim and to define in a dependent claim that the nail is a dowel or a hook-and-loop fastener. This leads to the consequence that the wording of the main claim and the parallel claims must be drafted very carefully. Furthermore, as a guideline, the dependent claims may contain specific features, while the main claim should rather comprise abstracted features. The example of the nail could therefore be included in a dependent claim, whereas "fastening means" or "means for fastening" would be suitable features for a main claim. A set of claims could therefore read as follows:

1. Device for …., comprising:
   a means for fastening the device to a wall, and
   …
4. Device according to one of the preceding claims, wherein the means for fastening is a nail, a dowel, or a hook-and-loop fastener.

In this example, it should be noted that the wording "means for fastening" must be used consistently. It is not advisable to describe a "fastening means" in another dependent claim, because from a patent law perspective, "means for fastening" and "fastening means" are considered different items.

9.14 The "Problem-Based" Claim Formulation

A "problem-based" claim formulation does not constitute a functional definition of a feature by means of a means-plus-function formulation. In a problem-based claim formulation, the features that lead to the solution of the invention's problem are deliberately omitted. Such a problem-based formulation is therefore not accepted in examination proceedings before a patent office. An example would be "an electric car with a high range." It may be an object within the claim to provide an electric car with a high range. It could also be that "an electric car with a high range" already refers to commercially available electric vehicles. In this sense, the feature would therefore either be "problem-based" and not executable[33] or unclear.[34] In both cases, the claim with this feature would be rejected by the patent office.[35]

9.15 Clarity

A claim must be executable. Terms such as "best possible," "optimal," "extreme," "maximum," "minimum," or verbs like "minimize," "maximize," "optimize," etc., are usually indeterminate terms. An exception exists, for example, if "maximum" and "minimum" are defined by the smallest and largest deflection of a mechanical structure. Otherwise, these terms are indeterminate and result in the scope of protection not being clearly defined. The terms "minimum," "maximum," "extremal," "best possible," "optimize," "minimize," and "maximize" should be avoided.

Article 84, sentence 2 of the EPC stipulates that claims must be "clear and concise." Therefore, in European proceedings, claims must be clearly formulated; otherwise, an application may be rejected. Great importance is attached to the clarity of claims in European proceedings, since only the claims are translated into all official languages, not the description.[36]

The German Patent Act does not contain a paragraph analogous to Article 84, sentence 2 of the EPC. Section 34, paragraph 4 of the Patent Act states that the invention must be described in the application "clearly and completely" so that it can be carried out by a person skilled in the art. Thus, the German Patent Act only requires that the application is clear so that the invention can be executed by the skilled person. However, there is no provision in the German Patent Act that requires the claims themselves, as opposed to just the application, to be clearly formulated.

Whether, nevertheless, a lack of clarity of the claims is a ground for rejection in the German grant procedure has not yet been definitively clarified by the highest courts.[37]

---

[33] Section 34 (4) Patent Act or Article 83 EPC.

[34] Article 84, sentence 2 EPC.

[35] EPO, Guidelines for Examination, Part F, Chapter IV, 4.10—Result to be Achieved, (retrieved on March 17, 2021, https://www.epo.org/law-practice/legal-texts/html/guidelines/d/f_iv_4_10.htm).

[36] EPO, Guidelines for Examination, Part F, Chapter IV, 4.2 Interpretation (retrieved April 5, 2021, https://www.epo.org/law-practice/legal-texts/html/guidelines/d/f_iv_4_2.htm).

[37] In an appeal procedure before the Federal Patent Court (15 W (pat) 9/13 "Polyurethane Foam"), the question of the clarity of the claims as a patentability requirement was addressed. The President of the German Patent and Trademark Office was involved in the proceedings, as it was a matter of fundamental importance. The President took the view that clarity was a requirement for patentability. The Federal Patent Court decided otherwise. An appeal on points of law was permitted but not filed by the President.

However, based on the current legal situation, it can be assumed that lack of clarity of the claims does not constitute a ground for refusal of a patent before the German Patent Office.[38]

Nevertheless, even for a German patent, the applicant should ensure that the claims are clear, as any lack of clarity will always be to the applicant's disadvantage. If a nullity action is brought against the patent and, due to the lack of clarity, a lack of novelty or inventive step is assumed, the patent will be revoked. Conversely, if in infringement proceedings, the lack of clarity of the claims means that infringement cannot be established, the lack of clarity will again prove disadvantageous for the applicant.

A lack of clarity does not arise merely because a single term has a broad meaning. If, in light of the context of the claim, it is clear to the skilled person what the scope of meaning of the term is, there is no lack of clarity.[39] Furthermore, the correct meaning of the term can be derived from the description of the patent, even in the case of a European patent.

9.15.1 Definitions

Unclear terms can be precisely defined by a definition in the description. However, care must be taken that the definition does not result in an overly narrow meaning of the defined term, which could unnecessarily reduce the scope of protection of a claim. In addition, the description should not contain contradictory definitions, as these would ultimately lead to ambiguity of the term.

9.15.2 Relative Terms

Terms such as "thin," "thick," "wide," "broad," "strong," "near," "adjacent to," "fast," "slow," "large," or "small" are so-called relative terms, which tend to be unclear. A definite meaning can only be derived from their context. Relative terms are therefore only permissible if, in the context of the relevant claim, a clear meaning for the respective relative term emerges. It is different if the term has a meaning accepted by experts. An example of this is the term "high frequency".[40]

---

[38] Federal Patent Court, 15 W (pat) 33/08 "Battery Monitoring Device," *Mitteilungen der Patentanwälte,* 2014, 126.

[39] Technical Board of Appeal EPO, T 456/91, November 3, 1993.

[40] EPO, Guidelines for Examination, Part F, Chapter IV, 4.6.1 (retrieved April 5, 2021, https://www.epo.org/law-practice/legal-texts/html/guidelines/d/f_iv_4_6_1.htm).

9.16 Special Terms

Terms are discussed that hold a particular significance in a claim.

9.16.1 "Device" and "Method"

The terms "device" and "method" can be used to characterize the subject matter of a claim. If there is a more precise designation for the subject matter, this can be chosen as an alternative. For example, if a harvesting machine is being claimed, the claim should not be formulated as "1. Device …" but rather as "1. Harvesting machine…". Alternatively, the terms "apparatus", "installation", or "means" can be used instead of "device". However, unless there are compelling arguments to the contrary, "device" should always be used instead of "means", "installation", or "apparatus", as the term "device" is always understood in a technical sense.

9.16.2 "consist of" versus "comprise" and "include"

The enumeration of features of a device or method can be made using the verbs "comprise/include" or "consist": "Device … comprising/including …" or "Device … consisting of …". When using the verb "consist", it should be borne in mind that the term is understood as exhaustive. In contrast, the terms "comprise" and "include" are not understood as exhaustive. This means that "comprise" and "include" do not result in a closed list. Advantageously, therefore, objects that possess all the features of a claim and, in addition, implement further features, are still to be regarded as infringing the claim. The situation is different with the term "consist", where the addition of a feature means that a potential competitor's product no longer falls within the scope of protection of the claim.

9.16.3 "bzw."

The term "beziehungsweise" has a multitude of meanings in German usage. For example, "bzw." can be understood as: and, or, furthermore, including, also, additionally, in addition, with, alongside, depending.[41] Due to this multitude of possible meanings, a claim can become unintelligible and its scope of protection cannot be determined precisely. The term "bzw." should therefore be avoided in a claim, but also in the description.

[41] Schulte/Moufang, Patent Act with EPC, Commentary, 10th edition, § 34 para. 137.

9.16.4 Vague terms such as "about" and "approximately"

It is understandable that the applicant wishes to broaden the scope of protection of a claim. However, it should be noted that a claim is interpreted from the perspective of a person skilled in the art. Deviations from the specifications of the claim, which the skilled person will also "read into" due to always present manufacturing tolerances, are included in the scope of protection. Vague terms such as "essentially", "approximately", or "about" are therefore usually superfluous and can even be detrimental, as they may lead to uncertainty regarding the scope of protection.

9.16.5 "in particular"

The terms "in particular", "preferably", or "for example" are used to provide an exemplary specification of a term. In a claim, it can be formulated: "… a vehicle, in particular a passenger car, …". This does not restrict the term "vehicle", but merely provides an example. This example may possibly be used during the grant procedure to distinguish from the prior art.

9.16.6 "at least", "at the very least", and "at a minimum"

In older patents, one often reads "…at least one…" or "…at least a…". This is intended to make clear that the feature may occur not just once, but any number of times in an embodiment of the invention. However, it should be noted that a claim only includes the absolutely necessary features. A plurality of a feature mentioned once is therefore always included in the scope of protection of the claim. The situation is different if not "one" or "a", but, for example, "2" or some other specific number is specified. In this case, it is assumed that exactly this number, namely 2 and not 3 or 4 or 5, is meant.

9.17 Suggestions from the Patent Office

A warning must be issued. The examiners at the patent office have as their first priority to prevent claims that are not legally valid. An examiner at the patent office will therefore tend to aim for a small scope of protection rather than the largest possible scope. The applicant, of course, has the opposite desire for a broad scope of protection in order to prevent competitors from circumventing the invention. A suggestion for an allowable claim from the examiner should therefore not be accepted without review. It may be a claim with a scope of protection that is too narrow.

Tip: Examine suggestions for claim wording from the patent office carefully. It is possible that a claim with too small a scope of protection is being proposed.

9.18 Case Study: Bag Combination for a Bicycle

Due to the recommended sequence in preparing a patent or utility model application, all aspects of the invention have already been addressed in the preceding sections of the application. There has been an intensive engagement with the subject matter of the invention. The invention is therefore present in all its details. Thus, the most challenging task in drafting an application can now begin, namely the formulation of the claims.

The purpose of the grant procedure is to determine the optimal, i.e., the broadest possible, scope of protection for the applicant. The broadest possible scope of protection is achieved if the claims are novel and involve an inventive step, and if nothing that is new and inventive remains unclaimed. The claims to be submitted with the patent application represent an initial attempt at formulation. However, it is advisable to make this initial attempt skillfully. Otherwise, significantly more responses to office actions will be required to obtain a patent. At the very least, the time until patent grant may be extended.

A small case study will be used to outline the approach to formulating claims. An invention could be a combination of bags for a bicycle:

> The invention is a bag combination consisting of two bags connected by a central piece. The central piece can be attached to a bicycle luggage rack. The bag can be removed from the luggage rack and taken into a changing room at a sports facility. The bag should be suitable for being hung in a locker. For this purpose, it has a hook. It must, of course, be ensured that even when hanging, the individual bags of the bag combination can be opened without all the contents falling to the floor. It should be considered that at least one bag of the bag combination is upside down when hanging. In particular, the bag combination therefore has at least one bag with a circumferential zipper. Furthermore, the bag combination can also have 4, 6, 8, or any number of bags.

The prior art consists of bicycle luggage rack bags with two bags. The features that are essential to the invention must be identified. Features are essential to the invention if the invention cannot be realized without them. These essential features are:

- Bag combination for the luggage rack of a bicycle. This is the prior art: this feature represents the generic term or preamble of a main claim.
- Bag combination with means for hanging in a locker. The main claim should not include the hook for hanging. The hook for hanging can be mentioned in a dependent claim. Including the hook would result in too narrow a scope of protection, as the bag combination could also be held in the locker by a magnet. Here, a "means-plus-function" formulation is appropriate, using the feature of a "means for hanging/attaching" or "attachment means."

In addition to essential features, there are features that lead to a particular embodiment of the invention. Such features further develop the invention and result in a specific form of the invention. These features belong in the dependent claims and under no circumstances in the independent claims.

If features for a particular embodiment of the invention are included in an independent claim, this immediately creates opportunities to circumvent the scope of protection of the patent. Nevertheless, it may be necessary to "elevate" a feature for a particular embodiment into the main claim in order to distinguish from a relevant prior art document.

The features of the dependent claims should have the potential that, by combining these features with those of the main claim, a claim results that is novel and involves an inventive step. Accordingly, dependent claims should not include technical trivialities that have no prospect of being granted.

A dependent claim should further develop the subject matter of the main claim. A dependent claim is linked to at least one feature of the main claim and develops it further. Therefore, a connection must be established between at least one feature of the dependent claim and at least one feature of the main claim, in particular by using an identical term. To make this possible, the terms in a set of claims must be used consistently.

Reference numerals are used in the claims, i.e., those reference numerals with which the individual elements are identified in the drawings. These reference numerals are placed in parentheses in the claims for better readability. Reference numerals are also used in the description of the drawings, but here the reference numerals are not placed in parentheses. This corresponds to the usual practice when drafting application documents. It is not permissible to refer directly to the drawings in the claims, for example by "... as shown in drawing Fig. 2 ..." or similar.[42] Claims should also not contain references to any other part of the application documents. The reason for this is that a claim should be understandable on its own. At most, the additional disclosure part of the application documents can be consulted for understanding.[43]

A set of claims could therefore read as follows:

1. Bag combination for arrangement on a luggage rack of a bicycle, comprising a bag and a means for fastening in a locker, wherein the means for fastening is arranged on the bag combination such that the bag combination can be hung in the locker.
2. Bag combination according to claim 1, further comprising 2, 4, 6, 8, or any number of separate bags and/or wherein the individual bags are connected by an intermediate piece and/or wherein the intermediate piece can be detachably connected to the luggage rack.
3. Bag combination according to claim 1 or 2, wherein at least one bag has a circumferential zipper.

[42] Section 9, paragraph 8 of the Patent Ordinance.

[43] Section 14, sentence 2 of the Patent Act or Article 69, paragraph 1, sentence 2 EPC.

9.19 Practical Examples with Recommendations

Below, original claims from real patent documents (granted patents, registered utility models, and published patent applications) are presented (in italics). Individual sections of the claims are shown in bold and underlined. These sections are identified as points ([Point 1], [Point 2], etc.). Comments follow for these points. Based on the notes on the claims, corrected claims are presented. Priority is given to examples from everyday life and bicycle technology, so as not to complicate the understanding of claim drafting with complex technical matters.

The first two examples contain no comments, as these claims are well formulated. They serve to provide a comparison with the following examples, which are less professionally formulated.

9.19.1 Example: Mirror with Display

The subject matter of DE 10 2015 104 437 B4 (Title: Mirror with Display) has the objective: "… to provide a device that, on the one hand, supports personal care, and on the other hand, offers a means of information transmission, in particular regarding physical condition, namely pulse." (DPMA, https://depatisnet.dpma.de/DepatisNet/depatisnet?action=pdf&docid=DE102015104437B4&xxxfull=1, accessed on 24 April 2026)

This objective is achieved by the main claim. The main claim reads:

"1. Mirror for reflecting a person, comprising:

a reflective surface (1), wherein the reflective surface (1) has a first section (21, 22, 23) that is at least partially transparent, and
a display (5) for displaying data,

characterized in that the

display (5) is arranged, from the perspective of the person, behind the first section (21, 22, 23), and wherein the data include information about the person's pulse, wherein the reflective surface (1) has a second section (20) that is at least partially transparent, wherein the mirror further comprises a device with a sensor in the form of an infrared sensor (2) for emitting infrared radiation to the person and for receiving the infrared radiation reflected from the person, wherein the sensor (2) is arranged, from the perspective of the person, behind the second section (20), wherein the mirror further comprises an interface (4) to additional devices, for example a blood pressure monitor, a heart rate monitor, a personal scale, and/or to the Internet, and wherein the interface (4) is a USB interface or a wireless interface, and wherein the mirror further comprises a control unit for controlling the display (5) and the infrared sensor (2) as well as for evaluating the reflected infrared radiation for the pulse." (DPMA, https://depatisnet.dpma.de/DepatisNet/depatisnet?action=pdf&docid=DE102015104437B4&xxxfull=1, accessed on 24 April 2026)

The main claim contains a very large number of features in the characterizing portion. As a result, the scope of protection is very limited. It is apparent that the applicant had to include a great many features in the main claim in order to obtain an allowable claim.

9.19.2 Example: End Mill

The subject matter of published application DE 10 2019 214 040 A1 (Title: End Mill and Method for Its Manufacture) is an end mill and a method for its manufacture. The invention is based on the objective of providing a milling tool suitable for milling a window. A further objective is to provide a manufacturing method for a milling tool according to the invention.

The objective is achieved by the main claim, which reads:

"1. Milling tool, in particular in the form of an end mill, with a cylindrical shank part having a central axis, to which a cylindrical cutting part (20) is connected, said cutting part having at least three helical peripheral cutting edges (22) separated from each other by flutes (24), which continue via cutting corner regions (26) into essentially radially extending end cutting edges (28, 30, 40), which, following the radially outer end cutting edge sections (28), each descend from the cutter end towards the central axis (AM) with a cutting edge section (30) formed by ground end pockets (32), characterized in that the cutting edge section (30) descends continuously up to the central axis (AM) and, in the region of the cutter core (36), is formed by a point (38) introduced into the end pocket (32), with which a center cutting edge (40) extending into the region near the central axis (AM) is produced." (DPMA, https://depatisnet.dpma.de/DepatisNet/depatisnet?action=pdf&docid=DE102019214040A1&xxxfull=1, accessed on 24 April 2026)

9.19.3 Example: Pressure Measuring Device

In the published patent application DE 10 2018 005 146 A1 (Title: Pressure Measuring Device), a pressure measuring device for a bicycle is described (see Fig. 9.5).

The main claim reads as follows:

"1. Pressure measuring device for a bicycle, wherein the pressure measuring device comprises:

a housing [Item 1]*;*
a pressure chamber accommodated in the housing;
a valve arranged in a first opening of the pressure chamber, wherein ***the valve is operatively configured to add or remove a fluid*** [Item 2] *along a flow path between the first opening and* ***the second opening*** [Item 3] *of the pressure chamber* ***into or out of a tire arrangement volume of a tire assembly*** [Item 4] *; and*

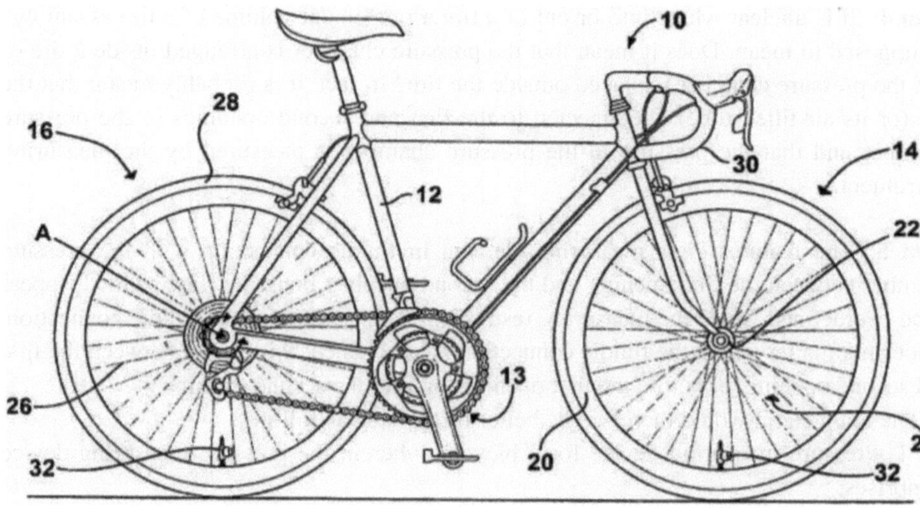

FIG. 1

Fig. 9.5 Pressure measuring device (DE 10 2018 005 146 A1)

a measuring element in fluidic connection with the pressure chamber ***between the first opening and the second opening along the flow path.*** [Item 5]." (DPMA, https://depatisnet.dpma.de/DepatisNet/depatisnet?action=pdf&docid=DE102018005146A1&xxxfull=1, accessed on 24 April 2026)

Item 1: The feature "a housing" is certainly not an essential feature of a pressure measuring device. However, this feature significantly limits the scope of protection, since the pressure measuring device could alternatively be arranged within the frame of the bicycle. The average skilled person would not consider the frame to be a housing of the subject matter of the invention. In this way, the scope of the main claim could be circumvented. This feature is therefore superfluous for the description of the invention and leads to an unnecessary restriction of the scope of protection of the main claim.

Item 2: The word "operatively" is superfluous. The clause "wherein the valve is operatively configured to add or remove a fluid …" could be more elegantly phrased as: "the valve for adding or removing a fluid…"

Item 3: A definite article is used for the second opening, although the term "second opening" has not yet been introduced. It should therefore read: "along a flow path between the first opening and **a second opening** of the pressure chamber."

Item 4: It is unclear what "into or out of a tire arrangement volume of a tire assembly" is supposed to mean. Does it mean that the pressure chamber is arranged inside a tire or that the pressure chamber is placed outside the tire? In fact, it is probably meant that the tire (or its air-filled tube) is connected to the first and second openings of the pressure chamber and that the pressure in the pressure chamber is measured by the measuring instrument.

Item 5: The features "...a measuring element in fluidic connection with the pressure chamber between the first opening and the second opening along the flow path..." appear to be overdefined, since the measuring instrument simply requires the fluidic connection. It does not matter where the fluidic connection is established; this can be between the first and second opening, or before the first opening, or after the second opening.

The main claim would therefore be better formulated as follows:

"1. Pressure measuring device for a bicycle, wherein the pressure measuring device comprises:

a pressure chamber with a first and a second opening;
a valve for adding or removing a fluid, which is arranged in a first opening of the pressure chamber; and
a measuring element in fluidic connection with the pressure chamber."

9.19.4 Example: Weather Protection Device

The document DE 10 2013 105 423 A1 (Title: Weather protection device for detachable mounting on a two- or multi-wheeled vehicle) addresses the technical problem of providing a weather protection device for a two-wheeled vehicle (bicycle or motorcycle) that does not impair the riding characteristics of the two-wheeler (see Figs. 9.6 and 9.7).

The main claim reads:

"1. Weather protection device (1) for detachable mounting on a two- or multi-wheeled vehicle ***consisting of*** [Point 1]

- ***at least*** [Point 2] *one flexible and at least partially transparent tarpaulin element (2)*
- ***at least*** [Point 2] ***one frame element between which the tarpaulin element (2) is arranged*** [Point 3] *and*
- ***at least*** [Point 2] *one fastening element for detachably mounting the weather protection device on a two- or multi-wheeled vehicle,*

characterized in that

- *an outer, endlessly continuous frame element (7) formed as a first frame (3)* ***consisting of*** [Point 1] *a flexible material accommodates the tarpaulin element (2) and*

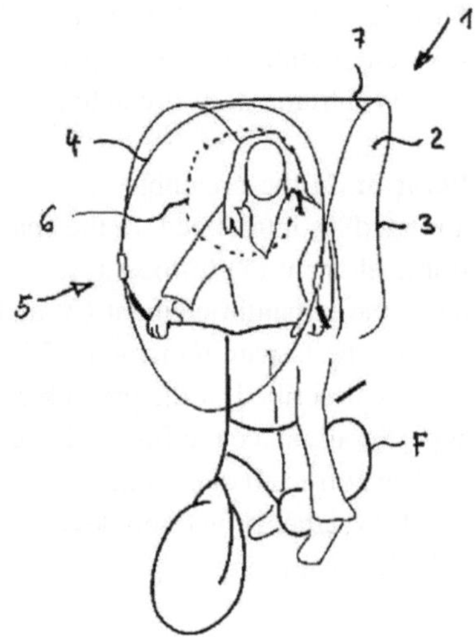

Fig. 9.6 Weather protection device Fig. 1 (DE 10 2013 105 423 A1)

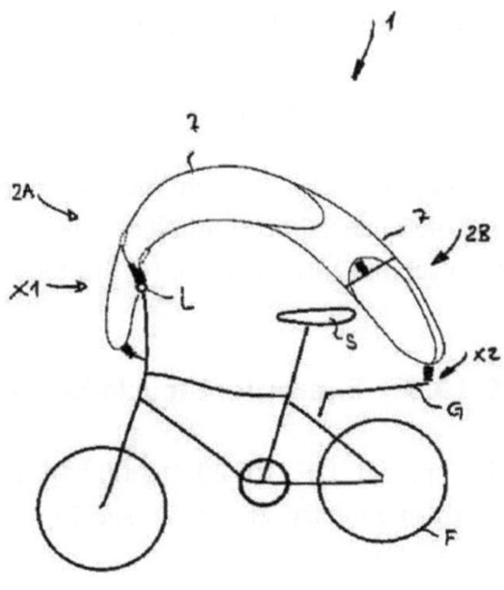

Fig. 9.7 Weather protection device Fig. 2 (DE 10 2013 105 423 A1)

– ***at least*** [Point 2] *a second further frame (4) stretches a partial area of the tarpaulin element (2)." (DPMA, https://depatisnet.dpma.de/DepatisNet/depatisnet?action=pdf&docid=DE102013105423A1&xxxfull=1, accessed on 24 April 2026)*

Point 1: Here, the word "consist" was used. "Consist" is interpreted as exhaustive, so that the invention contains exclusively the features following "consist". The invention cannot have any additional features (otherwise, instead of "consist," one would have had to use "comprise" or "include"). This formulation has the consequence that all weather protection devices which, although they have all the features of the main claim, also have additional features, no longer fall within the scope of protection of the main claim.

Point 2: The words "at least one" can be replaced by "one". "One" determines 1, 2, 3, 4, or any number of items. (Caution with the formulation "two frame parts." In this case, exactly 2 frame parts are determined, and not 1 or 3 or 4 frame parts.)

Point 3: In the preamble, a "frame element between which the tarpaulin element (2) is arranged" is introduced. In the characterizing part, it states "an outer, endlessly continuous frame element (7) formed as a first frame (3) consisting of a flexible material accommodates the tarpaulin element (2) and at least a second further frame (4) stretches a partial area of the tarpaulin element (2)". The question arises as to why the frame element was already described in the preamble. The description of the feature in the preamble appears superfluous, since the frame element is explained in detail in the characterizing part.

The corrected claim reads:

"1. Weather protection device (1) for detachable mounting on a two- or multi-wheeled vehicle comprising:

- a flexible and at least partially transparent tarpaulin element (2) and
- a fastening element for detachably mounting the weather protection device on a two- or multi-wheeled vehicle,

characterized in that

- an outer, endlessly continuous frame element (7) formed as a first frame (3) of a flexible material accommodates the tarpaulin element (2) and
- a second further frame (4) stretches a partial area of the tarpaulin element (2)."

9.19.5 Example: Bicycle Luggage Rack

The patent DE 10 2020 000 514 B3 (Title: Universal Bicycle Luggage Rack) was filed on January 28, 2020 (see Fig. 9.8). The publication date of granting a patent was

Fig. 9.8 Bicycle luggage rack (DE 10 2020 000 514 B3)

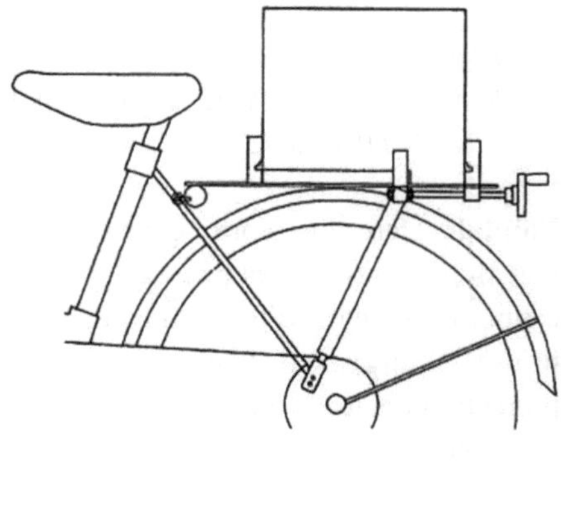

March 25, 2021. The application was granted with the main claim as originally filed. No amendments to the main claim were necessary. The patent was therefore granted suspiciously quickly.

The main claim reads:

*"1. The universal bicycle luggage rack is characterized in that it securely transports various containers with basic dimensions of 220*400 to 320*400 without slipping, and can be mounted on most bicycles with always the same parts in place of the previous luggage rack, wherein a frame made of high-strength flat material is formed around the basic dimension a=210 and b=373 and c=210, in which a and c are perpendicular to b and thus form a 3-sided frame; and wherein, to form the side-4 frame, flat high-strength material is bent at the ends at right angles so that a width of 400 mm is created and the edges provide flush support for the cargo; and wherein, as a frame carrier, a high-strength flat piece is used on which the 3-sided frame and the side-4 frame are mounted; and wherein 2 high-strength square blocks connect the frame of the universal bicycle luggage rack to the frame carrier; and wherein the front square block, the front cuboid, is screwed centrally above the frame carrier to it and to the 3-sided frame, and grooves are milled into it that correspond to the recesses of various box types to prevent vertical movement of these cargo items; and wherein the middle square block, the middle holder, is screwed below the frame carrier to it and to the side-4 frame, and a bore is provided centrally in the direction of travel for the sliding rod, in which it is rotatably mounted, and threads are tapped on the sides in the direction of travel to accommodate the retaining rods; and wherein a third square block, the rear cuboid, provided with grooves like those of the front cuboid, is slidably mounted on the frame carrier, in which a slot with the dimensions of the frame carrier is milled, through which the carrier is inserted, and wherein a thread for the sliding rod is tapped into the rear cuboid, by which the rear cuboid is moved so that transport boxes of different lengths sit flush in the frame; and wherein a handwheel or rotary knob is attached to the rear end of the sliding rod to manually push the rear cuboid flush against the transport box; and wherein 3 high-strength discs, the retaining discs, serve as mounts between the bicycle frame and the universal bicycle luggage rack, and wherein a thread, the central thread, is tapped into the center of the retaining discs; and wherein, in one of the retaining discs, the carrier retaining disc, two further threads are tapped parallel and radial to the axis and are thus screwed to the frame carrier of the universal bicycle luggage rack; and wherein, in two of the retaining discs, the bicycle retaining discs, near the edge of one side and at half the radius and transverse to the cut surface, another thread, the external thread, is tapped, into which eye bolts are screwed to provide attachment to the bicycle at the top; and wherein, for attachment to the Pletscher plate, the bicycle retaining discs with the external thread facing inward are screwed onto the disc rod and rest flush against the carrier retaining disc and can thus be screwed with the external thread to the two holes in the Pletscher plate spaced 1 1/8 inches apart; and wherein, for attachment to the seat stays, the bicycle retaining discs with the external thread facing outward are screwed onto the disc rod so that the external threads with the screwed-in ring bolts are bolted*

to the seat stay mounts with the ring, and the bicycle retaining discs rest on the seat stays; and wherein two strip bolts and two mounting tubes connect the lower part of the bicycle frame to the universal bicycle luggage rack; wherein each steel strip is welded centrally on its narrow side to a threaded rod, and wherein there is a hole at the bottom of the strip for possible mounting on the quick-release axle and above it a thread is tapped to be fastened to a mounting hole on the down tube of the bicycle or at the drop-out." (DPMA, https://depatisnet.dpma.de/DepatisNet/depatisnet?action=pdf&docid=DE102020000514B3&xxxfull=1, accessed on 24 April 2026)

The main claim is the only claim in the patent specification. There are no independent claims that additionally define a scope of protection. The main claim of the patent specification contains a large number of features and describes a specific bicycle luggage rack in great detail. It should also be noted that the infringement courts (special regional and higher regional courts, the so-called patent litigation chambers) only reluctant accept so-called equivalent patent infringement. Equivalent patent infringement exists if at least one feature of the main claim is not identically fulfilled. In this case, the differing feature must have the same technical effect and be of equal value. Furthermore, it must not require inventive activity to arrive at the equivalent feature. Until around 2002, equivalent patent infringement was generously accepted by the infringement courts. Currently, only literal infringements are accepted.[44] This means in this case that an infringement only exists if all features of the main claim are fulfilled. A potential infringer therefore only needs to implement one feature differently, and does not even have to be inventive, to obtain a workaround. In addition, they can benefit from the know-how disclosed in the patent specification. The patent holder has thus disclosed their know-how and, in return, has not received sufficient protection for their invention. This example confirms the saying "a well-written patent protects, a poorly written patent harms."

9.19.6 Example: Light or Line Laser Beam

The technical teaching of document DE 20 2019 002 356 U1 (Title: Light or line laser beam mounted laterally on a bicycle as a distance indicator for other vehicles during overtaking) aims to "increase road safety, both for the cyclist themselves and for other road users" (see Fig. 9.9).

The main claim of the application reads:

"1. Light beam or line laser beam (1) laterally on the bicycle, ***to indicate the designated minimum distance to the bicycle during lateral overtaking and to improve the***

[44] BGH, X ZR 16/09 "Occlusion device", *Industrial Property Protection and Copyright,* GRUR, 2011, 701; BGH, X ZR 76/14 "V-shaped guide arrangement", *Industrial Property Protection and Copyright,* GRUR, 2016, 1254.

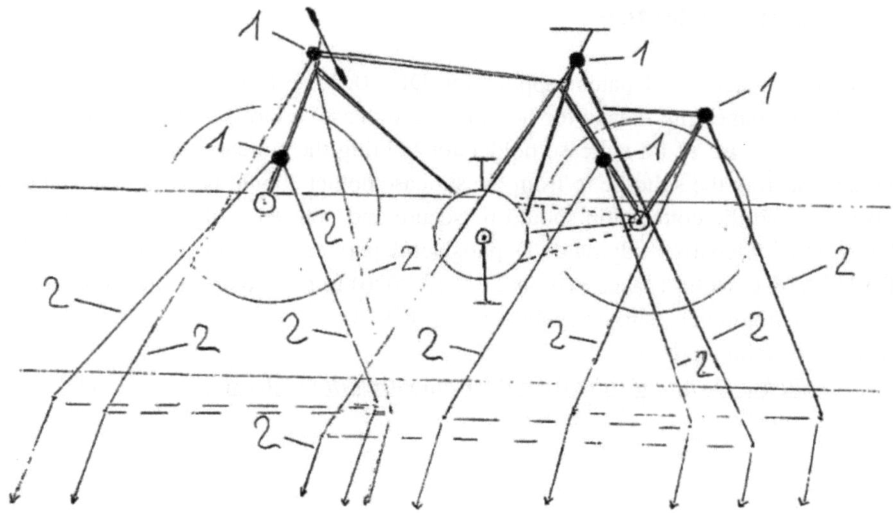

Fig. 9.9 Light and line laser beam (DE 20 2019 002 356 U1)

visibility of the bicycle at road intersections when turning right or left [Point 1], *characterized in that the light beam or line laser beam (1)* ***is mounted laterally*** [Point 2], *so that it* ***projects at a certain angle*** [Point 3] *parallel onto the roadway." (DPMA,* https://depatisnet.dpma.de/DepatisNet/depatisnet?action=pdf&docid=DE202019002356U1&xxxfull=1, *accessed on 24 April 2026)*

Point 1: The words "to indicate the designated minimum distance to the bicycle during lateral overtaking and to improve the visibility of the bicycle at road intersections when turning right or left" describe the objective to be achieved. These words do not specify which features are required to realize the invention. This passage can therefore be omitted.

Point 2: The preamble already states that the laser is arranged laterally on the bicycle. The feature "is mounted laterally" in the characterizing part is therefore redundant.

Point 3: For the feature "certain angle", the question arises as to the specific angle, since some angle must be chosen for the laser to shine onto the roadway. The feature of a "certain angle" is therefore indeterminate and unnecessary.

The corrected claim therefore reads:

"1. Bicycle with a device for generating a light beam or line laser beam (1), characterized in that the light beam or line laser beam (1) projects parallel to the bicycle onto the roadway."

9.19.7 Example: Suitcase

The technical teaching of patent application DE 10 2017 129 157 A1 (Title: Suitcase with rail-like connecting means for holding the suitcase) addresses the objective of "… providing a suitcase and a suitcase holder for holding the suitcase, wherein connecting and disconnecting the suitcase to/from the suitcase holder should be as simple and quick as possible, and the connection should be secure and tailored to the needs of the user." (DPMA, https://depatisnet.dpma.de/DepatisNet/depatisnet?action=pdf&docid=DE102017129157A1&xxxfull=1, accessed on 24 April 2026) In particular, the attachment of the suitcase to a motorcycle is to be enabled (see Fig. 9.10).

The main claim reads:

"1. ***A suitcase (1) and a suitcase holder matching the suitcase*** [Point 1], *wherein*

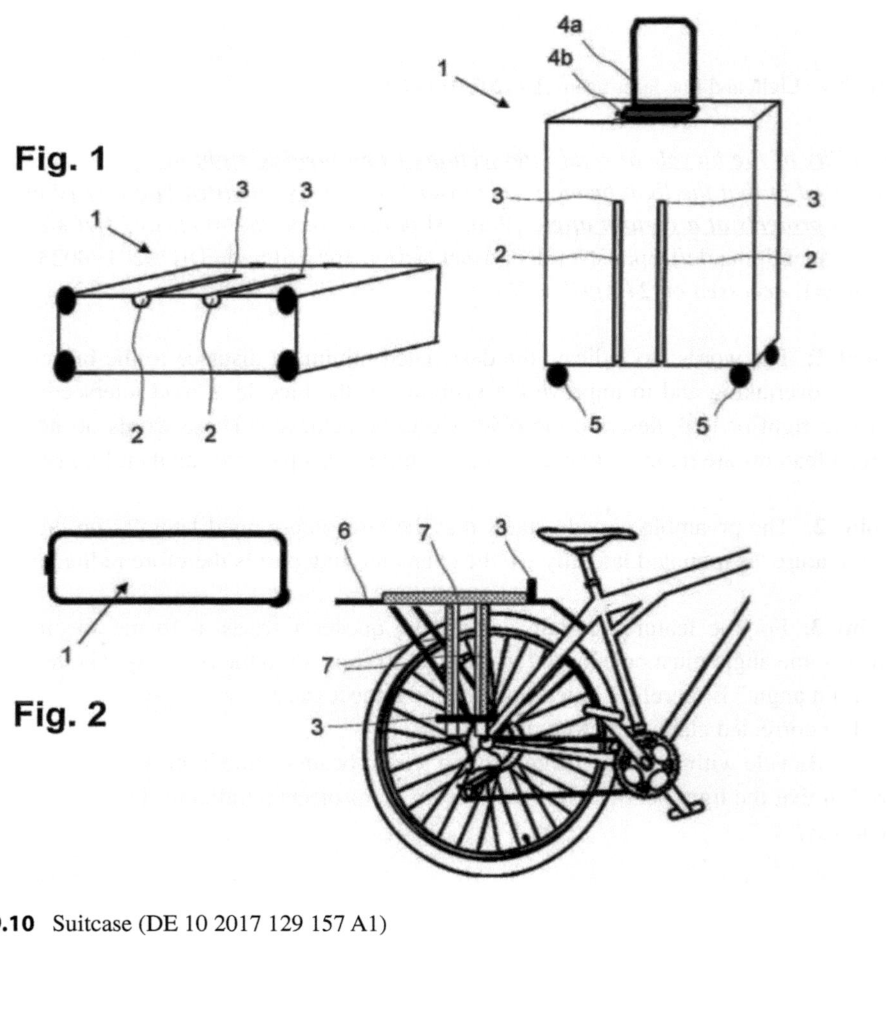

Fig. 9.10 Suitcase (DE 10 2017 129 157 A1)

a) *the suitcase* ***comprises an upper carrying handle wall, an opposite bottom wall, and four side walls in between*** [Point 2], *wherein on* ***at least*** [Point 3] *one of the side walls a first connecting means (2) is arranged and rigidly connected to the side wall,* ***to hold and transport the suitcase (1) at the first connecting means (2)*** [Point 4]*;*
b) *the suitcase holder has a second connecting means (7), which can be manually connected to and disconnected from the first connecting means (2),*

characterized in that

c) *the first connecting means (2) comprises at least one outer sliding rail or inner sliding rail in the longitudinal direction vertically between the carrying handle wall and the bottom wall;*
d) *the second connecting means (7) comprises at least one inner sliding rail or outer sliding rail in the longitudinal direction, each of which* ***communicates*** [Point 5] *with the respective outer or inner sliding rail of the first connecting means (2) and is designed such that they can be inserted into each other at a respective insertion position and slide into each other along a common longitudinal axis with only one degree of freedom along the longitudinal axis up to an end position;*
e) ***wherein the respective outer sliding rail has a C-profile with a cavity having a cavity width and a slot open laterally outward, wherein the slot is narrower than the cavity width;***
f) ***wherein the respective inner sliding rail has an inner sliding rail profile matching the cavity; and***
g) ***at least one stop (3) is arranged on the first connecting means (2) or on the second connecting means (7), which is designed, to stop the respective other first (2) or second connecting means (7) at the end position, so that the suitcase is held in the end position.*** [Point 6]" *(DPMA, https://depatisnet.dpma.de/DepatisNet/depatisnet?action=pdf&docid=DE102017129157A1&xxxfull=1, accessed on 24 April 2026)*

Point 1: In the present claim, two objects are claimed that can interact with each other. In this case, it is appropriate to refer to a system. The words "a suitcase holder matching the suitcase" are superfluous, because what does it mean for the suitcase holder to match the suitcase? The complementary interaction is explained by subsequent features, so the words "matching the suitcase" can be omitted.

Point 2: A suitcase always has an upper carrying handle wall, an opposite bottom wall, and four side walls in between. These features are therefore superfluous.

Point 3: "at least" is superfluous.

Point 4: The claim states that "on **at least** one of the side walls a first connecting means (2) is arranged and rigidly connected to the side wall". The question then arises as to what the subsequent feature "to hold and transport the suitcase (1) at the first connecting means (2)" is intended to achieve, since "holding" is already accomplished by the connecting means being attached to the side wall. "Transporting" is a functional formulation that can be omitted.

Point 5: "communicates" is not a technical term. "Complementarily formed" would perhaps be a more suitable formulation.

Point 6: This is followed by a concrete description of a specific embodiment, which should not be included in a first claim of a main claim, but rather in a dependent claim.

Corrected main claim:

1. System comprising a suitcase (1) and a suitcase holder, wherein

a) a first connecting means (2) is arranged on a side wall of the suitcase and rigidly connected to the side wall, and
b) the suitcase holder has a second connecting means (7), which can be detachably connected to the first connecting means (2),

characterized in that

c) the first connecting means (2) comprises at least one outer sliding rail or inner sliding rail in the longitudinal direction vertically between the carrying handle wall and the bottom wall;
d) the second connecting means (7) comprises at least one inner sliding rail or outer sliding rail in the longitudinal direction, each of which is complementarily formed with the respective outer or inner sliding rail of the first connecting means (2), so that they can be inserted into each other at a respective insertion position and slide into each other along a common longitudinal axis with only one degree of freedom along the longitudinal axis up to an end position.

9.19.8 Example: Bicycle with Arm and Leg Drive

The technical teaching of application DE 10 2014 015 824 A1 (Title: Bicycle with Arm and Leg Drive) fulfills the objective that a "cyclist can optimally transfer his driving force to the bicycle and at the same time steer the bicycle safely and precisely" (see Fig. 9.11, DPMA,).

The main claim reads:

"1. Bicycle ***with two wheels*** [Point 1] *and with arm drive,* ***in particular for carrying out the method according to one of claims 6 to 8*** [Point 2],

characterized in that, that

FIG.1

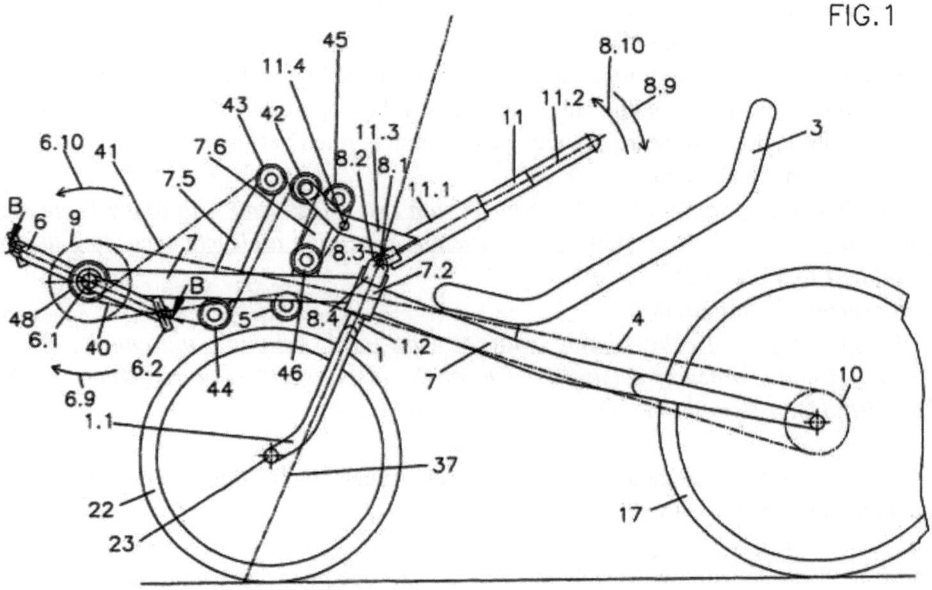

Fig. 9.11 Bicycle with arm and leg drive (DE 10 2014 015 824 A1)

the bicycle ***at least*** [Point 3] *has a pivotable handlebar (11, 18)." (DPMA, https://depatisnet.dpma.de/DepatisNet/depatisnet?action=pdf&docid=DE102014015824A1&xxxfull=1, accessed on 24 April 2026)*

Point 1: A bicycle usually has two wheels. However, there are also bicycles with three wheels. Tricycles are most likely not intended to be excluded from the scope of protection, so the words "with two wheels" can be omitted.

Point 2: Features introduced with "in particular" represent exemplary embodiments and are initially disregarded. A reference to subsequent claims is not permissible. A claim may only contain a reference to preceding claims.

Point 3: "at least" is superfluous. "A" or "one" is sufficient. With "a" or "one," two, three, four, five, or any number of handlebars are always included.

The corrected claim, as it would be interpreted by a patent office examiner, therefore reads:

"1. Bicycle with arm drive,

characterized in that, that

the bicycle has a pivotable handlebar (11, 18)."

It is unclear what an arm drive means, as no features are provided in the claim to inform the skilled person. Otherwise, it is a normal bicycle, since a handlebar is always pivotable. Possibly, the additional drive is to be achieved by pivoting. However, this cannot be inferred from the claim. This claim misses the opportunity for a good start in an official examination procedure. A quick patent grant is ruled out.

A parallel method claim is:

*"6. **Method for riding a bicycle*** [Point 1] ***with two wheels and with at least one pivotable handlebar (11, 18)*** [Point 2], ***in particular for riding a bicycle according to one of claims 1 to 5*** [Point 3],

characterized in that, that

the cyclist propels the bicycle with at least arm swinging movements [Point 4]." *(DPMA, https://depatisnet.dpma.de/DepatisNet/depatisnet?action=pdf&docid=DE102014015824A1&xxxfull=1, accessed on 24 April 2026)*

Point 1: What does "method for riding a bicycle" mean? Isn't riding the bicycle already the method? Nonsensical wording.

Point 2: "Bicycle with two wheels and with at least one pivotable handlebar": Typically, a bicycle has two wheels and a pivotable handlebar. These words can be omitted.

Point 3: Features introduced with "in particular" are initially disregarded.

Point 4: "the cyclist propels the bicycle with at least arm swinging movements": How does the cyclist do that? No features are specified that fulfill this objective (propelling the bicycle by arm swinging movements).

Corrected parallel claim:

6. Riding a bicycle according to one of claims 1 to 5.

This results in a completely meaningless method claim. It will be difficult for the applicant to "get back on track" in an examination procedure.

9.19.9 Example: Bicycle Dynamo

The utility model specification DE 20 2005 011 231 U1 (Title: Bicycle Dynamo, attachable and detachable from the bicycle without tools) contains the following protection claim[45]:

*1. Bicycle dynamo, **attachable and detachable from the bicycle without tools*** [Point 1], *characterized in that **this bicycle dynamo can be attached to the bicycle fork or. to the bicycle frame*** [Point 2] ***by means of a pre-installed bracket*** [Point 3] ***and can be***

[45] For a utility model, one does not speak of patent claims, but of protection claims.

attached and detached without tools [Point 4]. *(DPMA, https://depatisnet.dpma.de/DepatisNet/depatisnet?action=pdf&docid=DE202005011231U1&xxxfull=1, accessed on 24 April 2026)*

Point 1: "Attachable and detachable from the bicycle without tools": Only the objective is stated. The features by which the objective can be realized are missing. As an inventor, one should be especially critical and detached when drafting claims. Does the description actually cover what constitutes the invention? Are the features specified by which the invention can be realized, or only the advantages or even just the objective?

Point 2: A bicycle dynamo is always attached to the bicycle, in particular to the frame or the bicycle fork. This feature is obvious and also prior art and therefore superfluous.

Point 3: "by means of a pre-installed bracket": Obviously, the bicycle dynamo is attached to the bicycle with a bracket, and this can be pre-installed. These features of the main claim are also prior art and therefore superfluous.

Point 4: "can be attached and detached without tools": This is a formulation stating the objective. The features for realizing the invention are missing. In addition, these objective features are already present in the preamble. This passage can be deleted without replacement.

Corrected protection claim:

1. Bicycle dynamo.

The word "bicycle dynamo" obviously does not constitute a new or inventive protection claim. An obviously completely meaningless protection claim from which the applicant can derive no benefit.

9.19.10 Example: Dirt Cover for Bicycle

In the published patent application DE 19610832 A1 (Title: Dirt Cover for Bicycle), a device is presented into which a dirty bicycle can be packed in order to prevent contamination caused by the bicycle. Such contamination can occur, for example, during transport in the trunk of a vehicle. Furthermore, this device is intended to have only a small volume when not in use.

The patent application comprises a set of nine claims. The first six claims are as follows:

1. Dirt cover ***for a bicycle*** [Point 1], *characterized in that* ***it*** [Point 2] ***is made of flexible material and*** [Point 3] *can be closed by means of a closure flap (4)* ***can be closed*** [Point 4]. *(DPMA, https://depatisnet.dpma.de/DepatisNet/depatisnet?action=pdf&docid=DE000019610832A1&xxxfull=1, accessed on 24 April 2026)*

Point 1: "Dirt cover **for a bicycle**": It is often useful to clarify, using a phrase such as "for" or "to," what the device or the subject matter of the claim is intended for or what it is supposed to achieve. A better alternative would therefore be: Dirt cover for preventing contamination caused by a dirty bicycle. The phrase following "for" or "to" does not limit the scope of protection, as it merely clarifies what the subject matter of the claim is particularly suitable for. The fundamental application in other situations is not thereby excluded.

Point 2: With terms such as "this," "these," "it," "she," or "he," care must be taken to ensure it is clear which object is meant. It is therefore recommended to repeat "the dirt cover" instead of using "it" to avoid ambiguities.

Point 3: "is made of flexible material": This feature defines that the dirt cover is flexible over its entire surface, since the dirt cover is made of flexible material, i.e., consists only of flexible material. However, this is not required for the function of the dirt cover, as it only needs to be flexible in the sections where it is wrapped around the bicycle. The main claim therefore immediately allows for a workaround, namely a dirt cover that is not completely flexible. Therefore, instead of the verb "consist", "comprise" or "include" should be used to allow for areas of the dirt cover that are not made of flexible material.

Point 4: "can be closed": alternative wording: "closable"

The corrected claim reads:

1. Dirt cover for preventing contamination caused by a dirty bicycle, **characterized in that**

the dirt cover comprises flexible material and is closable by means of a closure flap (4).

The subject matter of the main claim is therefore a tarpaulin into which a bicycle can be wrapped, with an edge of the tarpaulin being used as a closure flap, since the edge is folded over the wrapped bicycle. The subject matter of the main claim is therefore fulfilled by any tarpaulin. The main claim describes a simple tarpaulin of the prior art that is large enough to wrap a bicycle. The main claim is therefore neither novel nor inventive and thus not legally valid.

2. Dirt cover for a bicycle according to claim 1, characterized in that by ***folding and joining the edges, a rectangular insertion pocket (9) is formed*** [Point 5]. (DPMA, https://depatisnet.dpma.de/DepatisNet/depatisnet?action=pdf&docid=DE000019610832A1&xxxfull=1, accessed on 24 April 2026)

Point 5: Folding and joining the edges of a tarpaulin to enclose or insert something is a typical application of a tarpaulin. Claim 2 is not a meaningful fallback position.

3. Dirt cover for a bicycle according to claim 1 or 2, characterized in that ***in the upper area*** [Point 6] ***a partial surface projects as a closure flap (4)*** [Point 7]. *(DPMA, https://depatisnet.dpma.de/DepatisNet/depatisnet?action=pdf&docid=DE000019610832A1&xxxfull=1, accessed on 24 April 2026)*

Point 6: "in the upper area": What is the upper area of a tarpaulin? A tarpaulin consists of two surfaces of an elongated, thin material. Is the upper area meant to be one of the two surfaces, or a section within a surface? If a section within a surface is meant, the question arises as to which section is being described. This is an unclear feature.[46]

Point 7: "a partial surface projects as a closure flap (4)": With any tarpaulin, a partial surface can project as a closure flap. Prior art is being described.

Claim 3 is not a possible fallback position in a grant procedure, as it is unclear ("in the upper area") or describes prior art.

4. ***Dirt cover for a bicycle according to claim 1*****,** **2 or 3** [Point 8], *characterized in that* ***fastening elements are provided on the cover and the closure flap (7)*** [Point 9]. *(DPMA, https://depatisnet.dpma.de/DepatisNet/depatisnet?action=pdf&docid=DE000019610832A1&xxxfull=1, accessed on 24 April 2026)*

Point 8: instead of "according to claim 1, 2 or 3", better: "according to any of the preceding claims"

Point 9: "fastening elements are provided on the cover and the closure flap (7)": Here, a new term "cover" is introduced. It is assumed that the applicant means the dirt cover by "cover." However, this is not necessarily the case; it could also refer to another object. Claim 4 is therefore unclear. Assuming the cover is the dirt cover, a distinction is made between the dirt cover and the closure flap. In fact, however, the closure flap is a part of the dirt cover. Furthermore, it is unclear why fastening elements are to be arranged on both the cover and the closure flap. Do these separate fastening elements perhaps interact, or what function do the fastening elements on the dirt cover serve, and what function do those on the closure flap serve? The entire dependent claim is therefore unclear. It is assumed that the applicant meant a fastening element characterized by complementary elements, for example, a button and a slot for receiving and holding the button.

5. Dirt cover for a bicycle according to claim 1, 2, 3 or 4, characterized in that the ***closure elements (7)*** [Point 10] *consist of double-sided adhesive* ***tapes*****,** ***hook-and-loop materials*****,** ***snap fasteners or a zipper*** [Point 11] **can** [Point 12]. *(DPMA, https://depatisnet.dpma.de/DepatisNet/depatisnet?action=pdf&docid=DE000019610832A1&xxxfull=1, accessed on 24 April 2026)*

[46] § 34 paragraph 4 PatG: "The invention must be disclosed in the application in such a clear and complete manner that a person skilled in the art can carry it out." Substantively identical: Article 83 EPC.

Point 10: A new term "closure elements" is introduced. The term "closure elements" is used with the definite article "the". Therefore, "closure elements" refers to a previously used term. Possible references include the closure flap or the fastening elements. Another possibility is that the use of the definite article was a mistake and thus "closure elements" was introduced entirely anew. Assuming that closure elements refer back to something mentioned earlier, the term could refer to the closure flap of claims 1 to 4 or the fastening elements of claim 4. The closure flap is used in the singular, the closure elements in the plural. This aspect suggests that closure elements and fastening elements are synonyms. On the other hand, claim 5 refers to all preceding claims 1 to 4. The feature of the closure flap is present in claims 1 to 4. The fastening elements were only described in dependent claim 4. This aspect again suggests that the closure elements refer to the closure flap. Therefore, dependent claim 5 cannot be interpreted in such a way that a skilled person can clearly understand it.

Point 11: "Consist of adhesive tapes, hook-and-loop materials, snap fasteners, or a zipper": The enumeration with "or" in combination with the concluding "consist of" means that only one of adhesive tape, hook-and-loop material, snap fasteners, or a zipper is claimed. A workaround would therefore be a simultaneous arrangement of a hook-and-loop material and a snap fastener.

Point 12: "Can": The claims should not describe what may or could be, but rather what the subject matter of the invention actually is. The main claim should also describe what is, and not set out possible embodiments. An exception exists if different cases that may occur alternatively are addressed and it is described what should apply in each case. This can apply, for example, to a method claim.

6. Dirt cover for a bicycle according to claim 1, 2, 3, 4, or 5, characterized in that ***the material used is PVC, PE, cotton, or a blended fabric*** [Point 13]. *(DPMA, https://depatisnet.dpma.de/DepatisNet/depatisnet?action=pdf&docid=DE000019610832A1&xxxfull=1, accessed on 24 April 2026)*

Point 13: A dirt cover can be made of PVC, PE, cotton, or a blended fabric. This is already known and therefore neither new nor based on an inventive step. Dependent claim 6 can therefore be omitted.

Summary: The analysis of the set of claims has shown that there is no invention with a prospect of patent grant. This result can be considered valuable, as it can save the costs and time involved in an official examination procedure. At the very least, annual fees can be avoided, which would be wasted here.

9.19.11 Example: Assembly Aid for a Bicycle

The technical teaching of published patent application DE 19609598 A1 (Title: Assembly Aid for a Bicycle) has the objective "to create a bicycle repair and maintenance system that, in a private setting, ensures a tip- and sway-proof as well as essentially rigid positioning of a bicycle, and at the same time can be stowed away in a standby position with just a few simple steps when not in use" (see Figs. 9.12 and 9.13, DPMA, https://depatisnet.dpma.de/DepatisNet/depatisnet?action=pdf&docid=DE000019609598A1&xxxfull=1, accessed on 24 April 2026).

Figure 9.12 (Fig. 3 of the published application) shows a side view of the assembly aid with the bicycle 2, an outrigger 7, a support tube 5, with the support tube 5 resting on the floor 3 after being swung down. The assembly aid is mounted on the wall 4.

The main claim reads:

1. Assembly aid (1) for a bicycle (2) with ***holding means (6)*** [Point 1] *for receiving the bicycle at a distance from the floor (3)* ***during a repair, maintenance*** [Point 2] ***or the like*** [Point 3], *characterized by a* ***by means of*** [Point 4] *an outrigger (7) to be pivotably mounted on a wall (4)* ***optionally*** [Point 5] *a support tube (5) to be placed on the floor (3) or, by pivoting the outrigger (7), to be placed against the wall (4),* ***on the side of which facing away from the wall (4) the holding means (6) are provided*** [Point 6]. *(DPMA, https://depatisnet.dpma.de/DepatisNet/depatisnet?action=pdf&docid=DE000019609598A1&xxxfull=1, accessed on 24 April 2026)*

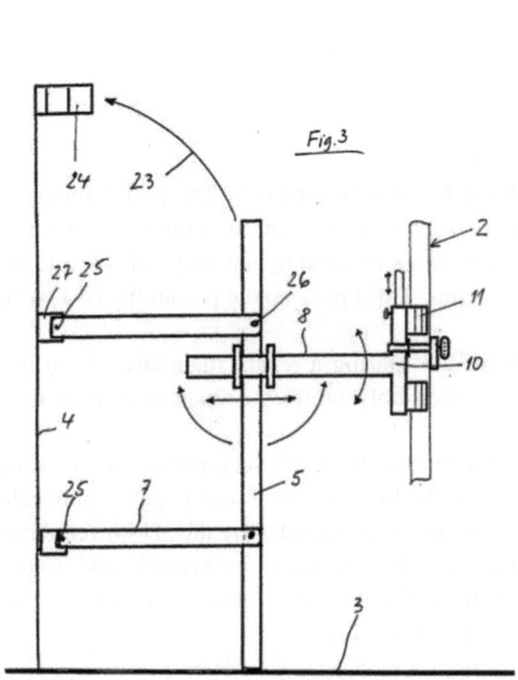

Fig. 9.12 Assembly aid for a bicycle Fig. 1 (DE 19609598 A1)

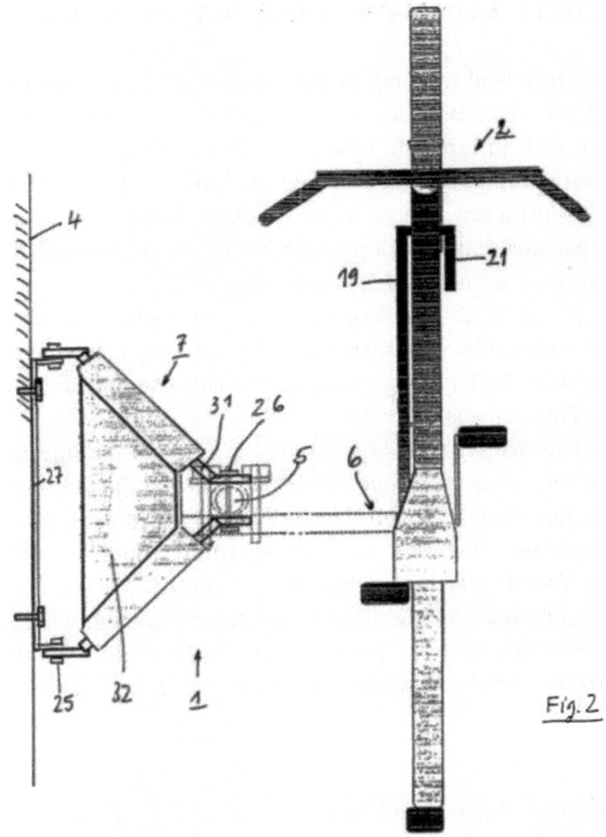

Fig. 9.13 Assembly aid for a bicycle Fig. 2 (DE 19609598 A1)

Point 1: "Holding means": The plural should only be used if the relevant feature must be present 2, 3, 4, 5, or any number of times. An assembly aid with only one holding means is not covered by the scope of protection, even though this variant is shown in the drawings. This opens up the possibility of a workaround solution by an imitator.

Point 2: "During a repair, maintenance": It would be better to use a formulation with "for": assembly aid for the repair or maintenance of a bicycle.

Point 3: "Or the like": Such phrases as "or similar", "etc.", "and the like", "and others", and "or the like" are often seen in patent applications. The applicant wishes to have objects of a similar type included by this. However, these phrases are of no value, as the applicant must describe the invention precisely and accurately. Features that are only described as "or the like" are not considered to be described precisely and therefore do not belong to the scope of protection.

Point 4: "By means of": an unnecessary phrase

Point 5: "Optionally": Unnecessary wording; instead, alternatives can be connected with "or".

Point 6: "On the side facing away from the wall (4) the holding means (6) are provided": It would also be possible to support the bicycle between the holding means and the wall. This feature excludes this technical solution and thus represents a workaround for an imitator. Therefore, this feature should not be in the main claim, but rather described in a dependent claim.

The corrected main claim reads:

1. Assembly aid (1) for the repair or maintenance of a bicycle (2) with a holding means (6) for receiving the bicycle at a distance from the floor (3),

characterized by

an outrigger to be pivotably mounted on a wall (4) for pivoting and placing a support tube on the floor (3) or against the wall (4).

9.19.12 Example: Automatic Control of Functions on Bicycles

The published patent application DE 10 2013 013 406 A1 (Title: Automatic Control of Functions on Bicycles by Means of a Multifunctional Grip) describes the control of functions on bicycles using a multifunctional grip. The technical objective is to provide a multifunctional grip with which several operationally appropriate functions can be performed with a single grip movement, e.g., braking, brake light adjustment, lowering the saddle for a secure stand, and engaging the starting gear (see Fig. 9.14, DPMA, https://depatisnet.dpma.de/DepatisNet/depatisnet?action=pdf&docid=DE102013013406A1&xxxfull=1, accessed on 24 April 2026).

The main claim of the application reads:

1. ***Automatic control of functions on bicycles*** [Item 1] ***without and with motor assistance*** [Item 2], ***during riding operation, as well as in holding and pushing mode*** [Item 3] *by means of a grip* ***1*** [Item 4], ***(multifunctional grip, actuating element)*** [Item 5] *with switching device 2* ***for at least two functions*** [Item 6], ***retrofit or integrated*** [Item 7], ***which*** [Item 8] *is attached to the handlebar or frame 4 by means of a fastening element 3, characterized in that the switching device 2, by actuating the grip 1, can trigger and cancel at least two different,* ***operationally appropriate, sequentially or simultaneously executed*** [Item 9] ***functions on bicycles*** [Item 10]. *(DPMA, https://depatisnet.dpma.de/DepatisNet/depatisnet?action=pdf&docid=DE102013013406A1&xxxfull=1, accessed on 24 April 2026)*

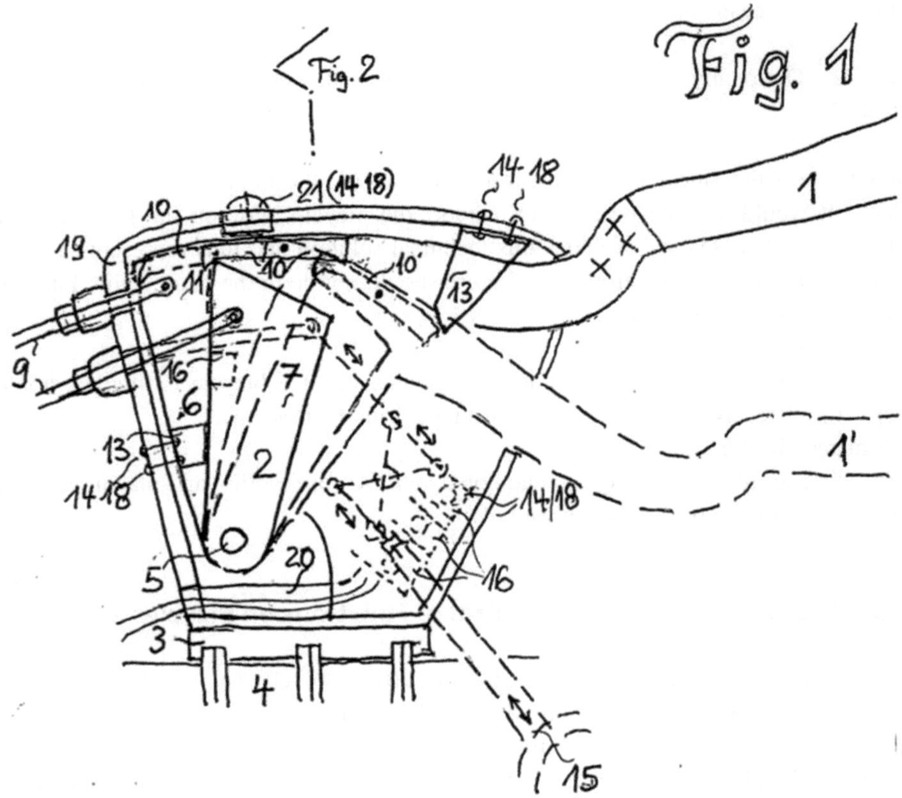

Fig. 9.14 Automatic control of functions on bicycles (DE 10 2013 013 406 A1)

Item 1: "Automatic control of functions on bicycles": The designation of the subject matter of the main claim misses the opportunity to give the invention a clear name. An example would be: "Control device for a bicycle."

Item 2: "without and with motor assistance": This feature says nothing, as it describes all possible situations. The feature can be omitted.

Item 3: "during riding operation, as well as in holding and pushing mode": This feature states that the control device can always be used. The feature can be omitted, as it always applies.

Item 4: Reference numerals in the claims are placed in parentheses.

Item 5: "(multifunctional grip, actuating element)": In a claim, an item should be designated by a single term, and not by several terms for the same item. Such multiple naming of an object does not clarify the invention.

Item 6: "for at least two functions": meaningless feature. Whether the control device has one or more functions cannot be a relevant feature. It would be much more important to specify which function or functions are implemented and how. This feature can be omitted.

Item 7: "retrofit or integrated": So the control device is prepared for any situation. In this case, this feature is meaningless and can be omitted.

Item 8: "which": "which" probably refers to the grip. To avoid ambiguities, it would be better to actually name the grip.

Item 9: "operationally appropriate, sequentially or simultaneously executed": Again, every situation is described. This feature can be omitted. It may indeed be the case that the feature, that a device performs a function in every state, is a special feature that establishes novelty and inventive step. In other cases, however, such features do not provide any means of distinguishing from the prior art and can be omitted.

Item 10: "can trigger and cancel functions on bicycles": A control device has a function, but does not trigger or cancel a function.

The corrected claim therefore reads:

1. Control device for a bicycle with a grip (1) and a switching device (2), wherein the grip (1) is attached to the handlebar or frame (4) by means of a fastening element (3),

characterized in that

the switching device (2) can perform at least two functions by actuating the grip (1).

Such a control device can be found on a bicycle in a multitude of ways. For example, a bell has a grip that can transfer the bell as a control device into the function "ring" or "not ring." The main claim of the patent application therefore has no prospect of being granted. Thus, a promising start to the grant procedure was not achieved.

This application underwent a patent grant procedure and resulted in patent DE 10 2013 013 406 B4. The main claim of the patent reads:

"1. Device for the automatic control of the height adjustment of a saddle on a bicycle with and without motor assistance during riding operation, as well as in parking and pushing mode by means of the braking process, using a brake lever (1) and a gas pressure or other spring as a movable seat post, as well as by means of a switching device (2) on the brake lever (1), characterized in that the switching device (2) is equipped with a central axis (5), about which, in an additional plane, a lever 2 (7) with an actuating element for the height adjustment of the saddle is pivotable, wherein the brake lever (1), extended by a lever 1 (6) with an actuating element for the brake in a first plane, and in

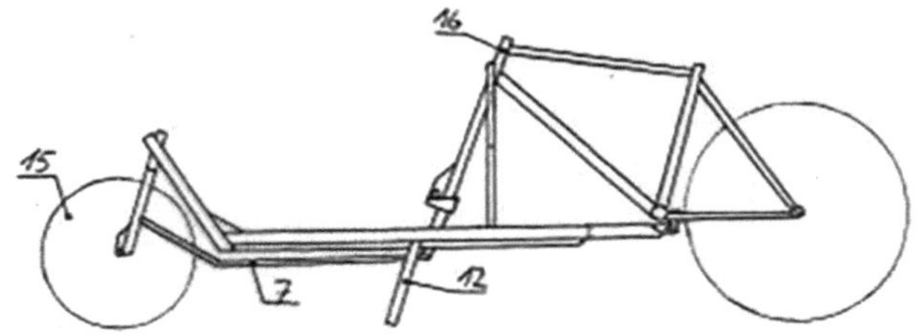

Fig. 9.15 Mountable loading platform Fig. 1 (DE 20 2020 002 966 U1)

the second plane, is provided with an adjustable driver (10) for the precise timing of the saddle lowering, wherein the adjustable driver (10) co-moves lever 2 (7) for the height adjustment of the saddle."[47]

The subject matter of the granted main claim has little in common with the main claim of the patent application. Moreover, it is a very specific subject matter with a very small scope of protection. It is questionable whether the patent has high economic value and can effectively prevent circumvention solutions.

9.19.13 Example: Mountable Loading Platform

In the utility model specification DE 20 2020 002 966 U1 (Title: Mountable loading platform with steering for bicycles (cargo bike)), the technical problem addressed is the conversion of a bicycle into a cargo bike (see Figs. 9.15 and 9.16, DPMA, https://depatisnet.dpma.de/DepatisNet/depatisnet?action=pdf&docid=DE202020002966U1&xxxfull=1, accessed on 24 April 2026).

This problem is solved by the main claim:

1. ***Connecting element (B1-B2)*** [Point 1], *characterized in that* ***the connecting element (B1) is permanently attached to the bicycle (C)*** [Point 2], ***positively engages*** [Point 3] *with the connecting element (B2) and by means of* ***the*** [Point 4] ***star knob screw (14)*** [Point 5] *can be screwed on and off* ***from the connecting element (B2)*** [Point 6]. [Point 7] *The connecting element (B2), in turn,* ***positively engages*** [Point 3] ***with the square tube of the loading platform (A) and is permanently connected to it*** [Point 8]. *(DPMA, https://depatisnet.dpma.de/DepatisNet/depatisnet?action=pdf&docid=DE202020002966U1&xxxfull=1, accessed on 24 April 2026)*

Point 1: "Connecting element (B1-B2)": Upon reviewing the claim, it becomes apparent that there are two connecting elements, namely B1 and B2. Therefore, "connecting element (B1-B2)" is incorrect. It would make more sense to refer to a system of two con-

[47] DE 10 2013 013 406 B4, DPMA, https://depatisnet.dpma.de/DepatisNet/depatisnet?action=pdf&docid=DE102013013406B4&xxxfull=1, accessed on 23 April 2026.

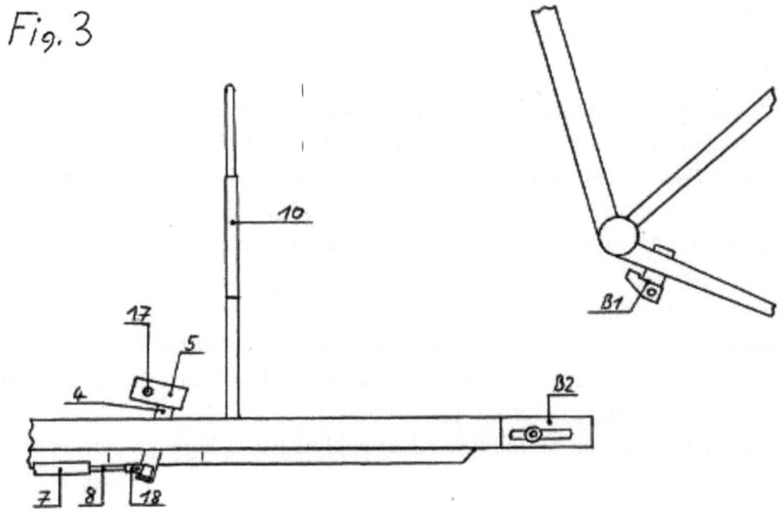

Fig. 9.16 Mountable loading platform Fig. 3 (DE 20 2020 002 966 U1)

necting elements (B1, B2) that can be detachably connected, with one connecting element (B1) arranged on a bicycle and the second connecting element (B2) on a loading platform for a cargo bike.

Point 2: "the connecting element (B1) is permanently attached to the bicycle (C)": The bicycle is introduced for the first time here, so it should be "to a bicycle" rather than "to the bicycle." A permanent attachment can be more elegantly expressed as "non-detachably connected."

Point 3: "positively engages": This is a very specific type of connection. A non-positive, i.e., frictional, connection would already constitute a workaround solution. Therefore, the positive connection should only be described in a dependent claim and not in the main claim.

Point 4: "by means of the star knob screw": A star knob screw is introduced for the first time here, so the indefinite article "a" must be used.

Point 5: The star knob screw is a very specific feature. Using a different screw would already provide a workaround solution for an imitator.

Point 6: Instead of "can be screwed on and off," the phrase "can be detachably connected" can be used.

Point 7: A claim must be exactly one sentence. A claim cannot consist of two or more sentences.

Point 8: "with the square tube of the loading platform (A) and is permanently connected to it": The loading platform is introduced for the first time here, so the indefinite article must be used. A permanent connection can be referred to as "non-detachably connected." The square tube is a very specific feature, which allows an imitator to circumvent the claim by using a tube with a different cross-sectional shape.

The corrected claim reads:

1. System of two connecting elements (B1, B2), which can be detachably connected, wherein one connecting element (B1) is arranged on a bicycle and the second connecting element (B2) on a loading platform for a cargo bike.

The corrected claim can be interpreted as applying to a luggage carrier of a bicycle and is therefore not legally valid.

9.19.14 Example: Small Bicycle Trailer

The utility model specification DE 20 2019 002 093 U1 (Title: Small bicycle trailer that can be placed on the luggage rack, folded down when needed, used, and folded back up) addresses the technical problem of providing a combination bicycle that features a trailer which can be placed on a luggage rack. This allows, in particular, grocery shopping to be done by bicycle (see Fig. 9.17, DPMA, https://depatisnet.dpma.de/DepatisNet/depatisnet?action=pdf&docid=DE202019002093U1&xxxfull=1, accessed on 24 April 2026).

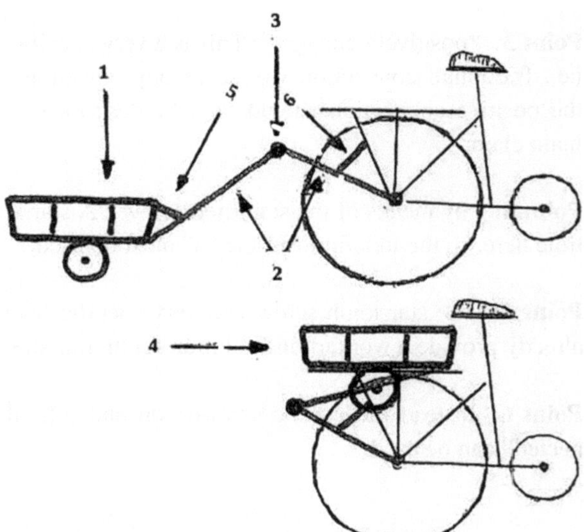

Fig. 9.17 Small bicycle trailer (DE 20 2019 002 093 U1)

The main claim reads:

1. ***Small*** [Point 1] *bicycle trailer* ***(load class presumably 50kg)*** [Point 2], *which can be placed permanently or temporarily on the luggage rack of a bicycle* ***and is suitable,*** [Point 3] *to be folded down when needed, used for transport tasks, and folded back up* ***(Fig. 1)*** [Point 4] , *characterized in that the* ***trailer*** [Point 6] *by means of a linkage on the trailer and on the bicycle frame* ***(Fig.2)*** [Point 5] *as well as a coupling* ***(Fig.3)*** [Point 5] *on a* ***standard*** [Point 7] *bicycle luggage rack* ***with clamping plate is securely carried during travel*** [Point 8], *and, if necessary,* ***can be set down on the ground and reattached with just a few simple steps*** [Point 9]. *(DPMA, https://depatisnet.dpma.de/DepatisNet/depatisnet?action=pdf&docid=DE202019002093U1&xxxfull=1, accessed on 24 April 2026)*

Point 1: "small": What does "small" mean, and what is then considered large? This is an unclear term, which leads to ambiguity in the claim.

Point 2: "(load class presumably 50 kg)": Apart from reference signs, nothing in a claim should be in parentheses, as parentheses indicate that something is not important. If something is not important, it can be omitted. This is especially true for the main claim, in which only those features should be listed that are essential for the realization of the invention. "Presumably" also does not belong in a claim, as only those features that are precisely defined are to be included in claims.

Point 3: "placed permanently or temporarily": This feature can be omitted, as the invention is obviously not dependent on it, since it appears suitable for any state.

Point 4: "to be folded down when needed, used for transport tasks, and folded back up": In a claim, the features should be described with which the invention can be realized. Therefore, it should be specified how the folding down and folding back up are to be carried out. Otherwise, only the tasks that the invention is supposed to accomplish are listed.

Point 5: "(Fig. 1)": In the claims, references to the drawings (or the description) should be avoided.[48]

Point 6: "trailer": It is to be assumed that "trailer" refers to the bicycle trailer. Terms should be used consistently in a set of claims to avoid allegations of lack of feasibility or ambiguity.

Point 7: "standard bicycle luggage rack": This implies that the use of the bicycle trailer with a non-standard, i.e., a rather rarely sold, bicycle luggage rack does not constitute an

[48] Section 9, paragraph 8 of the Patent Ordinance.

infringement of the utility model. This feature "standard" can dramatically reduce the scope of protection and play into the hands of an imitator.

Point 8: "securely carried during travel with clamping plate": There is a problem of lack of feasibility, as it is unclear where and how the clamping plate is to be arranged in order to achieve a "secure" arrangement for travel. Furthermore, the question arises as to when something is considered "secure for travel."

Point 9: "can be set down on the ground and reattached with just a few simple steps": This property, that something can be detached and reattached with a few simple steps (what is "few"? What is a "simple step"? Does one just grab, or also turn or push?), applies to any part of a bicycle. Therefore, this feature cannot serve to establish novelty and inventive step and can be omitted. In addition, it should be described how this task, i.e., setting down and reattaching with a few simple steps, is achieved.

The corrected claim reads:

1. Bicycle trailer, which can be arranged on the luggage rack of a bicycle, characterized in that the bicycle trailer is carried on the bicycle luggage rack by means of a linkage on the bicycle trailer and on the bicycle frame as well as a coupling.

The main claim describes a bicycle trailer that can be attached to the bicycle frame and to the bicycle luggage rack by means of a coupling. The question arises as to why the bicycle trailer is fixed to two points on the bicycle.

9.19.15 Example: Bicycle Seat Post Alarm System

The technical objective of the invention described in utility model DE 20 2010 011 175 U1 (Title: Bicycle Seat Post Alarm System) is to provide a device for a bicycle that prevents the theft of a bicycle equipped in this way (see Fig. 9.18, DPMA, https://depatisnet.dpma.de/DepatisNet/depatisnet?action=pdf&docid=DE202010011175U1&xxxfull=1, accessed on 24 April 2026).

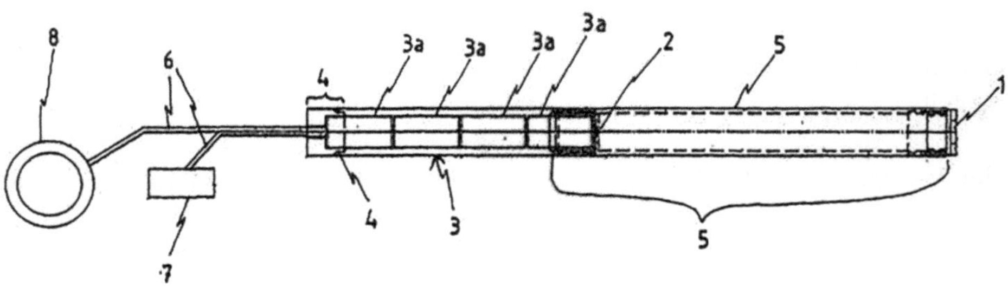

Fig. 9.18 Bicycle seat post alarm system (DE 20 2010 011 175 U1)

The utility model specification comprises two claims:

1. ***Arrangement of an alarm system*** [Point 1] ***on a two-wheeler*** [Point 2], *in particular on a bicycle, for preventing the theft of the two-wheeler, wherein the alarm system is contained in the seat post* ***or.*** [Point 3] *seat tube* ***contained*** [Point 4]. *(DPMA, https://depatisnet.dpma.de/DepatisNet/depatisnet?action=pdf&docid=DE202010011175U1&xxxfull=1, accessed on 24 April 2026)*

Point 1: The claim refers to the arrangement of an alarm system, i.e., moving an alarm system to a two-wheeler and connecting the alarm system to the bicycle. The main claim is therefore not a device claim, but a method claim. This was probably not intended by the applicant. The main claim should therefore preferably read "Alarm system on a two-wheeler."

Point 2: "on a two-wheeler" corresponds to "on the two-wheeler": The two-wheeler is newly introduced. Therefore, it must be accompanied by the indefinite article "a," and not the definite article "the".

Point 3: "or": The abbreviation "or" is linguistically ambiguous.

Point 4: "contained": A better formulation would be "arranged".

The corrected claim reads:

1. Alarm system on a two-wheeler, in particular on a bicycle, for preventing theft of a two-wheeler, wherein the alarm system is arranged in the seat post or seat tube.

The set of claims includes a second claim that does not refer to the preceding claim. The utility model specification therefore contains a main claim and an independent claim, each defining a separate scope of protection.

2. Arrangement of an alarm system on a two-wheeler, in particular on a bicycle, for the **tracking** (Point 1) of the theft of a two-wheeler, wherein **a** (Point 2) GPS receiver **and.** (Point 3) transmitter is contained in the seat post or seat tube. (DPMA, https://depatisnet.dpma.de/DepatisNet/depatisnet?action=pdf&docid=DE202010011175U1&xxxfull=1, accessed on 24 April 2026)

Point 1: "Tracking": "Tracking" may be a better choice of word than "determination."

Point 2: "a": A GPS receiver and transmitter is newly introduced, so the indefinite article must be used.

Point 3: "and.": No abbreviations in the claims! To avoid ambiguities, words must be written out in full. In a claim, every word counts. Therefore, one should not risk something being misunderstood due to an ambiguous abbreviation. A claim is not a normal continuous text, where incorrect or poorly chosen words can be corrected based on the (extensive) context. If a poorly chosen word contradicts the rest of the context, the claim becomes unclear and is therefore no longer allowable or legally valid. When drafting a claim, every word must be weighed carefully!

The corrected claim reads:

2. Alarm system on a two-wheeler, in particular on a bicycle, for tracking the theft of a two-wheeler, wherein a GPS receiver and transmitter is arranged in the seat post or seat tube.

9.19.16 Example: Theft Protection for Bicycles

The utility model specification DE 20 2012 010 902 U1 (Title: Theft protection for bicycles and other movable consumer goods) addresses as its technical objective the arrangement of a code on a bicycle, so that effective protection against bicycle theft is achieved and the recovery of a stolen bicycle is facilitated (see Fig. 9.19, DPMA, https://depatisnet.dpma.de/DepatisNet/depatisnet?action=pdf&docid=DE202012010902U1&xxxfull=1, accessed on 24 April 2026).

The main claim reads:

1. A theft protection device, characterized in that a reflective and signal-colored ***cuff*** [Point 1] *is* ***firmly adhered*** [Point 2] ***to*** [Point 3] *the product to be protected and contains an alphanumeric and/or QR code. (DPMA,* https://depatisnet.dpma.de/DepatisNet/depatisnet?action=pdf&docid=DE202012010902U1&xxxfull=1, *accessed on 24 April 2026)*

Fig. 9.19 Theft protection for bicycles (DE 20 2012 010 902 U1)

The arrangement of an alphanumeric or QR code that is reflective and in signal colors is certainly not new. This is especially true since the arrangement of the code is claimed for any products, not just bicycles.

Point 1: "Cuff": A simple sticker does not fully encircle a bicycle tube as a cuff does, and would be a workaround for an imitator.

Point 2: "Firmly adhered": What is strong, and what would be "weak" in comparison? Using vague terms endangers the required clarity of a claim.

Point 3: "Adhered": A connection with rivets would already be a workaround for an imitator.

Claims 2 and 3 read:

2. A theft protection device, characterized in that an app for smartphones with corresponding database connectivity has a "Stolen" and a "Found" button, which, when activated, automatically transmits the associated location data and user data to the database.

3. A theft protection device, characterized in that a database is configured so that all users of the associated app within a freely selectable radius from the theft location receive a corresponding push mail, if desired. (DPMA, https://depatisnet.dpma.de/DepatisNet/depatisnet?action=pdf&docid=DE202012010902U1&xxxfull=1, accessed on 24 April 2026)

It only becomes apparent when reading the second claim that this is a software application. In principle, technical software can be protected with a utility model. In this case, it would have been advisable to formulate the main claim accordingly to claim the software. The main claim, as it stands, is doomed to fail. It would therefore make sense to claim the software as a system. This would allow for a legally valid main claim. One possibility is:

1. System for protecting a bicycle against theft, comprising:

a reflective and signal-colored cuff that has an alphanumeric and/or QR code, wherein the cuff is arranged on a bicycle,
a recording device, in particular a smartphone, for reading the code, wherein the recording device transmits the code to software, and
the software, which identifies the bicycle based on the code.

2. System according to claim 1, wherein the cuff is permanently attached to the bicycle.

3. System according to one of claims 1 or 2, wherein the software, after identifying the bicycle, transmits a message to the owner of the bicycle.

4. System according to claim 3, wherein the message to the owner is sent according to adjustable criteria, in particular depending on the location of the owner.

5. System according to one of the preceding claims, wherein the recording device allows a choice between entering "Found" and "Stolen".

9.19.17 Example: Furniture for Storing a Bicycle

Utility model publication DE 20 2018 005 207 U1 (Title: Furniture for storage, presentation, repair, and cleaning of a bicycle, as well as for storing tools, clothing, accessories, and use-related items such as helmet, gloves, glasses, protective clothing, and the like) describes the technical task of providing a piece of furniture for an apartment that serves as a storage place for a bicycle. (DPMA, https://depatisnet.dpma.de/DepatisNet/depatisnet?action=pdf&docid=DE202018005207U1&xxxfull=1, accessed on 24 April 2026)

The set of claims reads as follows:

1. Bicycle furniture for ***neat and space-saving*** [Point 1] ***storage of bicycles, characterized in that, a bicycle*** [Point 2] *within* ***one's own apartment/house*** [Point 3] ***in a piece of furniture*** [Point 4] ***is to be fastened to a holding claw*** [Point 5]. *(DPMA, https://depatisnet.dpma.de/DepatisNet/depatisnet?action=pdf&docid=DE202018005207U1&xxxfull=1, accessed on 24 April 2026)*

Point 1: "Neat and space-saving": These words are task-related statements that do not belong in the claim, but rather in the description, to emphasize the inventive activity of the claim. Only those features by which the invention is realized, i.e., the features that ensure the bicycle can be stored neatly and in a space-saving manner in the apartment, are to be included in the claim.

Point 2: "Storage of bicycles, characterized in that a bicycle": It is elegant to introduce, in the first section of a claim, the objects on or with which the invention is realized. In this case, the storage of a bicycle can be described. "Storage of bicycles" should not be used, as otherwise the claim would refer to two or more bicycles and not also to bicycle furniture for just a single bicycle. This bicycle can then be further addressed, using the definite article. This clarifies with which object, here a bicycle, the invention can be realized and how this is to be done.

Point 3: "Within one's own apartment/house": Use in a hotel for a hotel guest would then, strictly speaking, not constitute an infringement of the property right. "Own" apartment is not a technical feature, but a legal one that defines the ownership relationship. The adjective "own" therefore does not belong in the claim. Only features that have a technical character and technically define the subject matter should be included in the claim.

Point 4: "In a piece of furniture": The indefinite article "a" introduces a new piece of furniture. However, it is to be assumed that the applicant meant the bicycle furniture by "a piece of furniture." Therefore, the definite article should be used. In addition, the scope of protection should not be drawn too narrowly. It is conceivable that the furniture only partially encloses the bicycle. The following wording would therefore be appropriate: "at least partially in the bicycle furniture". Terms should be used consistently. The term "bicycle furniture" should be used again, and not a new term "furniture" introduced.

Point 5: "Is to be fastened to a holding claw": This feature has no connection with the bicycle furniture, as it is not described that the holding claw is attached to the furniture. It would also be possible for the holding claw to be mounted on a wall. The holding claw is therefore not an essential element of the invention. It would also work without a holding claw. Additionally, the holding claw significantly limits the scope of protection, as another holding means could also be used, for example, a loop or a hook.

The feature of a "fastening means, for example a holding claw, a hook, or a loop", would therefore be well placed in a dependent claim.

The corrected main claim therefore reads:

1. Bicycle furniture for storing a bicycle, characterized in that the bicycle is accommodated at least partially within an apartment/house in the bicycle furniture.

9.19.18 Example: Bicycle with Rear Area Detection

The technical teaching of DE 20 2018 001 459 U1 (Title: Bicycle with Rear Area Detection) addresses the problem that cars rapidly approaching from behind pose a danger to cyclists. This danger is to be managed by a camera system. The technical objective of the invention is therefore to provide greater traffic safety for the cyclist, especially in cases where vehicles are approaching from behind. (DPMA, https://depatisnet.dpma.de/DepatisNet/depatisnet?action=pdf&docid=DE202018001459U1&xxxfull=1, accessed on 24 April 2026)

The main claim of the utility model reads as follows:

1. Bicycle with rear area monitoring, characterized in that ***on a bicycle*** [Point 1] ***one or more cameras*** [Point 2] ***are mounted in such a way, that the rear area of the cyclist is continuously captured by these*** [Point 3] *and* ***displayed on a monitor also mounted on the bicycle*** [Point 4] *in such a way that* ***the type of each vehicle, approaching the cyclist from behind, its speed, its distance from the cyclist, and its trajectory are recognizable, and by means of a distance measuring system also mounted on the bicycle, every vehicle approaching from behind is signaled to the cyclist in a perceptible manner*** [Point 5]. *(DPMA, https://depatisnet.dpma.de/DepatisNet/depatisnet?action=pdf&docid=DE202018001459U1&xxxfull=1, accessed on 24 April 2026)*

Point 1: "on a bicycle": The subject of the claim, the bicycle, has already been introduced (bicycle with rear area monitoring). For this reason, the definite article "on the bicycle" should be used here.

Point 2: "one or more cameras": Phrases such as "one or more", "at least one", or "a minimum of one" can be conceptually replaced with "one" or "a".

Point 3: "cameras, are mounted in such a way that the rear area of the cyclist is continuously captured by these": The word "continuously" should be removed from the main claim. Whether a camera records images intermittently or continuously does not affect the inventive step. The word "continuously" alone therefore does not contribute to inventive activity. On the other hand, the scope of protection is significantly restricted, as a camera that only records images intermittently would not fall within the scope of the main claim. The words "cameras, are mounted in such a way that the rear area of the cyclist is continuously captured by these" can be elegantly replaced by a "for" construction: "cameras for capturing the rear area of the cyclist".

Point 4: "displayed on a monitor also mounted on the bicycle": A camera system with a monitor arranged on the cyclist's helmet would not infringe the main claim as currently worded. On the other hand, this particular embodiment is likely the preferred one.

Point 5: "the type of each vehicle approaching the cyclist from behind, its speed, its distance from the cyclist, and its trajectory are recognizable, and by means of a distance measuring system also mounted on the bicycle, every vehicle approaching from behind is signaled to the cyclist in a perceptible manner": These features are very detailed, which opens the door to circumvention solutions for the main claim. These features should be included in the dependent claims so that, if necessary, individual features can be used to distinguish from the prior art.

The corrected main claim therefore reads as follows:

1. Bicycle with rear area monitoring, characterized in that a camera for capturing images of the rear area of the bicycle is arranged on the bicycle, wherein the images are presented on a monitor.

9.19.19 Example: Holder for Bicycle Front Lights

Document DE 20 2019 000 855 U1 (Title: Holder for Bicycle Front Lights for Optimizing the Light Emission Angle) describes the problem that the emission angle of a bicycle front light can only be adjusted when the bicycle is stationary, but that the emission angle changes when a rider mounts the bicycle (see Fig. 9.20). (DPMA, https://depatisnet.dpma.de/DepatisNet/depatisnet?action=pdf&docid=DE202019000855U1&xxxfull=1, accessed on 24 April 2026)

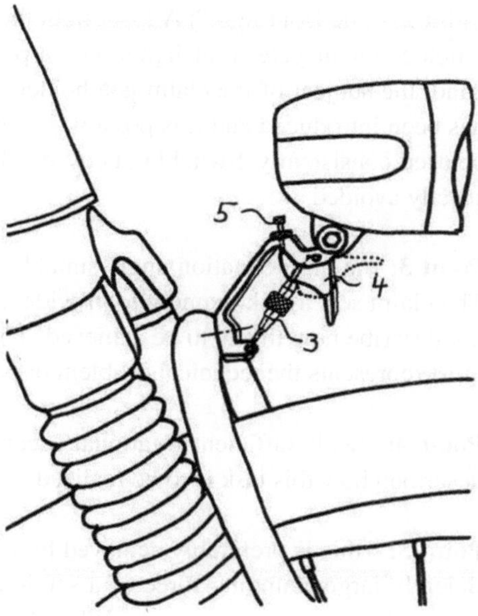

Fig. 9.20 Holder for bicycle front lights (DE 20 2019 000 855 U1)

The solution to this problem follows from the main claim, which reads:

1. Holder for ***bicycle front lights*** [Point 1] *for optimizing the light emission angle, characterized in that*

- ***the front light*** [Point 2] ***is connected to its holder at an inclination angle suitable for illuminating the roadway*** [Point 3] ***with sufficient rotational security*** [Point 4] ,
- ***this is preferably achieved by a positive fit, such as a square axle, or by a sufficiently large clamping force of a screw connection*** [Point 5] ,
- *and this* ***preset*** [Point 6] ***inclination angle of the front light*** [Point 7] *can be adjusted by a* ***control element almost uniformly steplessly, jerk-free and thus "precisely" or suitably graduated for roadway illumination, both with one hand, preferably without tools and thus possibly. during the ride, and can be corrected at any time by the cyclist according to his current needs for roadway illumination*** [Point 8] *. (DPMA, https://depatisnet.dpma.de/DepatisNet/depatisnet?action=pdf&docid=DE202019000855U1&xxxfull=1, accessed on 24 April 2026)*

Point 1: "Holder for bicycle front lights": a better formulation is "holder for a bicycle front light". This introduces the bicycle front light as a further subject, alongside the holder, and allows it to be referenced in the features of the main claim and the dependent claims.

Point 2: "the front light": A front light has not yet been introduced. The claim describes a holder for bicycle front lights, i.e., a plurality of bicycle front lights. If, on the other hand, the subject of the claim is a holder for a bicycle front light, the bicycle front light has been introduced and it is possible to refer to the bicycle front light. The terms should be used consistently. Switching between "bicycle front light" and "front light" should be strictly avoided.

Point 3: "at an inclination angle suitable for illuminating the roadway with its holder": The claim sets a task, namely to provide a "suitable" inclination angle. However, it does not describe how this is to be achieved. This passage does not belong in the claim, but at most represents the technical problem or part of the technical problem.

Point 4: "with sufficient rotational security": Again, only the task is set, but it is not described how this task is to be realized.

Point 5: "this is preferably achieved by a positive fit, such as a square axle, or by a sufficiently large clamping force of a screw connection": It is unclear how a positive fit is created by a single square axle. What does "sufficient" mean, when is a clamping force "sufficiently large"? This passage can be better formulated as: "in particular by a positive fit or a clamping force."

Point 6: "preset": Before which event or condition is something set? Can the inclination angle not also be "reset" and is the word "preset" therefore not superfluous?

Point 7: "inclination angle of the front light": The inclination angle has already been defined as the inclination angle of the front light. If the word "inclination angle" is used again, further specification is not necessary. It is sufficient to refer to the "inclination angle".

Point 8: "Control element almost uniformly stepless, jerk-free and thus 'precisely', or suitably graduated for roadway illumination, both with one hand, preferably without tools and thus possibly during the ride, can be corrected at any time by the cyclist according to his current needs for roadway illumination": Only task-oriented formulations are given, which do not describe how the invention is achieved, but only what the invention is supposed to accomplish. This description of the task belongs in the introductory part of the application, where the object of the invention is described.

The corrected claim reads:

1. Holder for a bicycle front light for optimizing the light emission angle, characterized in that the bicycle front light is connected to the holder at an inclination angle, in particular by a positive fit or a clamping force.

This main claim describes the prior art and is therefore not legally valid. Application of the main claim to infringement of intellectual property rights is not possible.

9.19.20 Example: Multipurpose Trailer for Bicycle and Boat

The utility model specification DE 20 2016 006 583 U1 (Title: Multipurpose trailer for bicycle and boat for mutual towing) addresses the technical problem of combining the recreational sports of cycling and boating. The invention provides a multipurpose trailer that can accommodate either a boat or a bicycle. When cycling, the boat is packed into the multipurpose trailer and towed by the bicycle. When boating, the multipurpose trailer holds the bicycle and is towed by the boat. An important feature of the multipurpose trailer is therefore its buoyancy (see Fig. 9.21). (DPMA, https://depatisnet.dpma.de/DepatisNet/depatisnet?action=pdf&docid=DE202016006583U1&xxxfull=1, accessed on 24 April 2026)

The main claim reads as follows:

1. ***The multipurpose trailer*** [Point 1] ***(Fig.1)*** *[Point 2]* ***with the following features*** [Point 3]*:*

a) ***Single- or multi-part*** [Point 4] *buoyant body with sufficient load capacity in water* ***for trailer*** [Point 5] *and bicycle* ***(Fig. 1–1)****,*
b) *Two or more wheels for travel on land* ***(Fig. 1–3)****,*
c) *Attachment option for* ***boats*** [Point 6] ***as trailer load*** [Point 7] *(Fig. 2),*
d) *Attachment option for* ***bicycles*** [Point 8] ***as trailer load*** *(Fig. 3),*
e) *Tow coupling for* ***bicycle as towing means*** [Point 9],
f) *Tow coupling for* ***boat as towing means*** [Point 10]. *(DPMA, https://depatisnet.dpma.de/DepatisNet/depatisnet?action=pdf&docid=DE202016006583U1&xxxfull=1, accessed on 24 April 2026)*

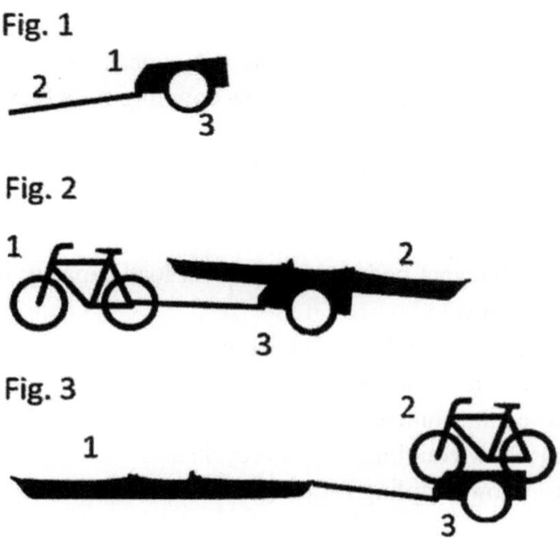

Fig. 9.21 Multipurpose trailer for bicycle and boat (DE 20 2016 006 583 U1)

Point 1: "The multipurpose trailer": No article is used in the designation of the subject matter of the claim. In particular, no definite article is used, as the subject matter has not yet been introduced.

Point 2: "(Fig. 1)": A reference to drawings or text passages in the description should be avoided. A claim should be understandable on its own.

Point 3: "with the following features": Only features are listed in a claim. This passage is superfluous and can be replaced by "comprising" or "including".

Point 4: "Single- or multi-part": Since there are no other variants than single- or multi-part, this passage does not define a specific feature. This statement can be omitted.

Point 5: "for trailer": It is assumed that "trailer" refers to the multipurpose trailer. The multipurpose trailer has already been introduced in the claim, so the definite article should be used here. Furthermore, terms should be used consistently. Therefore, the term "multipurpose trailer" should be used throughout, not just "trailer". This avoids ambiguities, which are always interpreted to the detriment of the applicant, since the applicant could have been explicit.

Point 6: "boats": It is more elegant to introduce the subject "boat", i.e., singular, so that this subject can subsequently be referenced and further specified.

Point 7: "as trailer load": superfluous

Point 8: "bicycles": It makes sense to refer again to the subject "bicycle", which has already been introduced, and then use the definite article.

Point 9: "as towing means": is superfluous, as it has already been described that the multipurpose trailer has a tow coupling for a bicycle.

The corrected main claim therefore reads as follows:

1. Multipurpose trailer comprising:

a) a buoyant body with sufficient load capacity in water for the multipurpose trailer and a bicycle,
b) two or more wheels for travel on land,
c) an attachment option for a boat,
d) an attachment option for the bicycle,
e) a tow coupling for the bicycle and
f) a tow coupling for the boat.

9.19.21 Example: Bicycle Trailer

Utility model publication DE 29822869 U1 (Title: Bicycle Trailer) sets out the objective of providing a bicycle trailer that can alternatively be used as a transport device without a bicycle, for example as a backpack or rolling suitcase (see Figs. 9.22 and 9.23, DPMA, https://depatisnet.dpma.de/DepatisNet/depatisnet?action=pdf&docid=DE000029822869U1&xxxfull=1, accessed on 24 April 2026).

The main claim reads:

1. Bicycle trailer with a running wheel (16), characterized by ***at least one*** [Point 1] ***backpack-like*** [Point 2] ***carrying device*** [Point 3] *(34). (DPMA, https://depatisnet.dpma.de/DepatisNet/depatisnet?action=pdf&docid=DE000029822869U1&xxxfull=1, accessed on 24 April 2026)*

Point 1: "at least one": can be replaced by "one".

Point 2: "backpack-like": This word has no clear meaning. It is therefore necessary to interpret the term "backpack-like". Here, backpack-like could mean that objects can be received and transported, or that the backpack-like object looks like a backpack, i.e., has pockets and comprises a curved side and a flat side. Furthermore, backpack-like could mean that the corresponding object can be used as a backpack by a person. In this case, a

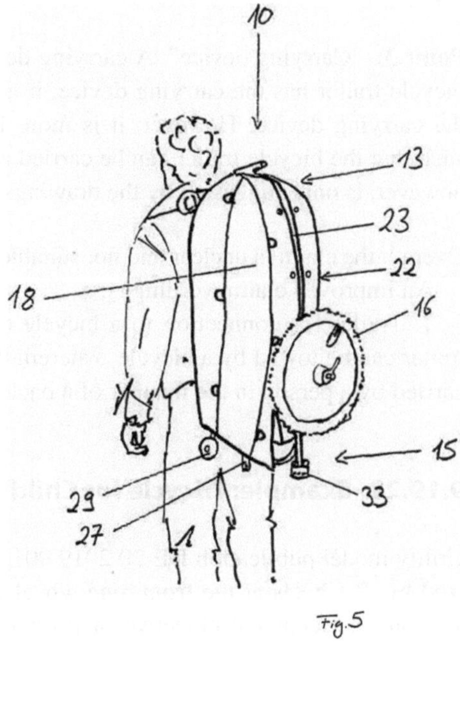

Fig. 9.22 Bicycle trailer Fig. 5 (DE 29822869 U1)

Fig. 9.23 Bicycle trailer
Fig. 6 (DE 29822869 U1)

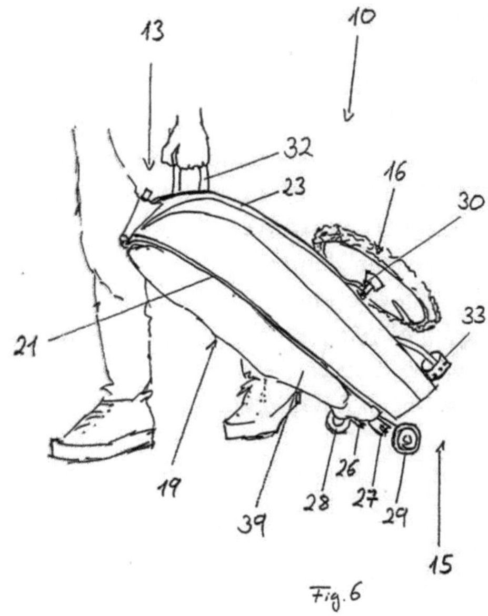

formulation such as "can be carried on the back by a person as a backpack" would have been possible. In summary, the term "backpack-like" cannot be assigned a clear meaning. The main claim is therefore unclear.

Point 3: "Carrying device": A carrying device carries something or is carried. Since the bicycle trailer has the carrying device, it is to be assumed that the bicycle trailer carries the carrying device. However, it is more likely meant here that the carrying device is such that the bicycle trailer can be carried as a backpack by a person. This interpretation, however, is only suggested by the drawings.

Overall, the claim is unclear and not suitable for defining the claimed intellectual property.

An improved claim wording is:

1. Trailer for connection to a bicycle comprising a running wheel (16), so that the trailer can be towed by a bicycle, wherein the trailer has a carrying strap so that it can be carried by a person in the manner of a backpack.

9.19.22 Example: Tricycle for Children

Utility model publication DE 20 2019 001 213 U1 (Title: Tricycle for children characterized by: 2 wheels at the front, one wheel at the rear, car steering wheel, coaster brake) sets out the technical objective of further developing a children's tricycle and, in particular, improving its steering characteristics. (DPMA, https://depatisnet.dpma.de/DepatisNet/depatisnet?action=pdf&docid=DE202019001213U1&xxxfull=1, accessed on 24 April 2026)

The set of claims consists of two claims:

1. Children's tricycle, characterized in that ***a steering wheel (like a car) is used*** [Point 1], ***instead of handlebars as on a bicycle*** [Point 2].

2. Children's tricycle, characterized in that ***the design provides*** [Point 3] *that* ***two 12-inch wheels are at the front and one 12-inch wheel at the rear*** [Point 4]. *(DPMA, https://depatisnet.dpma.de/DepatisNet/depatisnet?action=pdf&docid=DE202019001213U1&xxxfull=1, accessed on 24 April 2026)*

Point 1: "a steering wheel (like a car) is used": However, using a steering wheel like a car presupposes that a steering wheel is present. Therefore, the steering wheel should first be introduced as part of the children's tricycle. Parentheses should not be used in a claim, as parentheses indicate that something is less important. Less important features do not belong in a claim at all.

Point 2: "instead of handlebars as on a bicycle": Here it should be stated that the children's tricycle does not have a bicycle handlebar. In a claim, it should generally not be described what is not present, but rather what is present.

Point 3: "the design provides": How can a design, i.e., a thing and not a person, provide or plan something? Moreover, it is unclear what is meant by the design. If the design refers to the children's tricycle, then the children's tricycle should be mentioned instead of the design. This passage can be omitted.

Point 4: "two 12-inch wheels are at the front and one 12-inch wheel at the rear": Here, two features are mixed together. On the one hand, it is specified that the wheels have a diameter of 12 inches. In addition, it is described that the children's tricycle has two wheels at the front and one wheel at the rear.

The set of claims contains two claims that are formulated independently. There are therefore two claims, each with its own scope of protection. However, it is to be assumed that a dependency was intended, and that claim 2 is to be formulated as dependent on claim 1. The corrected set of claims reads:

1. Children's tricycle, characterized in that the children's tricycle has a steering wheel for steering the children's tricycle.

In the introductory part of the description, it can be explained that children's tricycles have so far had handlebars like those on a bicycle and that handlebars are disadvantageous compared to a car steering wheel. The disadvantage of the prior art can, in particular, be seen in the fact that a steering wheel enables more precise steering compared to the rough steering by means of handlebars.

2. Children's tricycle according to claim 1, characterized in that the children's tricycle has two wheels at the front and one wheel at the rear and/or that the wheels have a diameter of 12 inches.

9.19.23 Example: Sidecar Connection

In the utility model specification DE 20 2013 002 713 U1 (Title: Sidecar Connection for Bicycles), the technical problem addressed is that, due to a rigid connection of a sidecar to a bicycle, analogous tilting or leaning movements of the sidecar relative to the bicycle are prevented (see Fig. 9.24). (DPMA, https://depatisnet.dpma.de/DepatisNet/depatisnet?action=pdf&docid=DE202013002713U1&xxxfull=1, accessed on 24 April 2026)

The task description in document DE 20 2013 002 713 U1 is unfortunate, as the solution to the problem—i.e., the invention—is already anticipated in the statement of the problem. It is described that the problem arises due to a rigid connection between the bicycle and the sidecar. However, the step from "rigid" to "flexible" is a small one, which means that, in this presentation, the invention appears not to be based on inventive

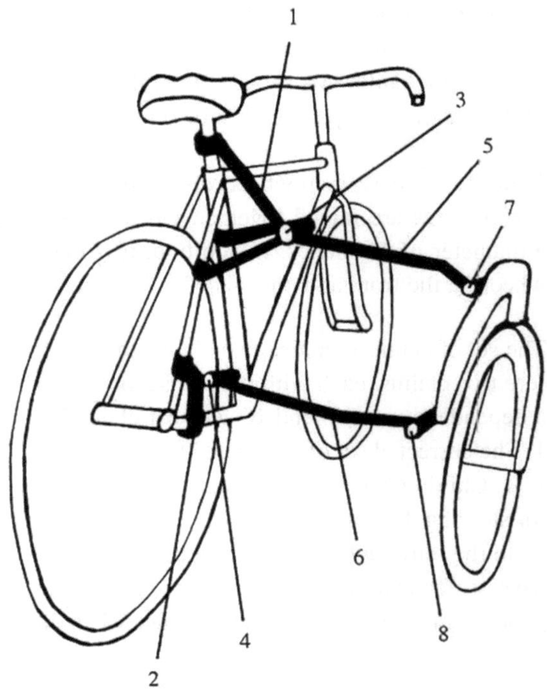

Fig. 9.24 Sidecar connection (DE 20 2013 002 713 U1)

activity. However, an obvious invention does not lead to a legally valid utility model.[49] Any reference to the rigid connection should be avoided in the statement of the problem.

The invention is realized by the main claim:

1. ***Leaning-enabled*** [Point 1] *bicycle-sidecar connection* ***(Fig. 1)*** [Point 2] ***for attaching sidecars*** [Point 3] *to* ***commercially available*** [Point 4] *bicycle frames, characterized in that* ***by means of the connection attached to the bicycle (Fig. 4, Fig. 5) and to the sidecar (Fig. 7, Fig. 9), quadruple bearing (Fig. 6, Fig. 7, Fig. 8, Fig. 9) of the upper and lower swing arms (Fig. 2, Fig. 3), the bicycle-sidecar combination can perform a parallel lean according to the pendulum principle*** [Point 5]. *(DPMA, https://depatisnet.dpma.de/DepatisNet/depatisnet?action=pdf&docid=DE202013002713U1&xxxfull=1, accessed on 24 April 2026)*

Point 1: "Leaning-enabled": By creating an adjective, it is described what the bicycle-sidecar connection is supposed to achieve. This adjective is task-oriented and therefore superfluous. Instead, the claim should describe the features by which the invention is realized.

Point 2: "(Fig. 1)": References to drawings and parts of the description should not be made in a claim.[50]

Point 3: "Bicycle-sidecar connection for attaching sidecars": Unnecessary duplication, since "bicycle-sidecar connection" already describes that the subject of the claim is a connection between a bicycle and a sidecar.

Point 4: "Commercially available": What is commercially available and what is not? What is the intended meaning of this feature? Is it even a technical feature? If not, it does not belong in the claim. If it is, it leads to a restriction of the scope of protection that the applicant is unlikely to have intended. In this case, an imitator would simply have to use a non-commercially available, i.e., rarely sold, bicycle frame to create a workaround.

Point 5: What are the upper and lower swing arms? What is a fixed bearing? What does an unfixed bearing look like? What is the bicycle-sidecar combination? Does this refer to the bicycle with the sidecar attached to it? What is a parallel lean? What is the pendulum principle? The characterizing part of the claim is unclear.

A corrected claim could read as follows:

1. Bicycle-sidecar connection, characterized in that the connection of the sidecar to the bicycle is effected by a bearing, so that the sidecar is pivotable relative to the bicycle.

The embodiment with four bearings can be described in a dependent claim.

---

[49] Section 4, sentence 1 of the Patent Act or Article 56, sentence 1 of the EPC.

[50] Section 9, paragraph 8 of the Patent Ordinance.

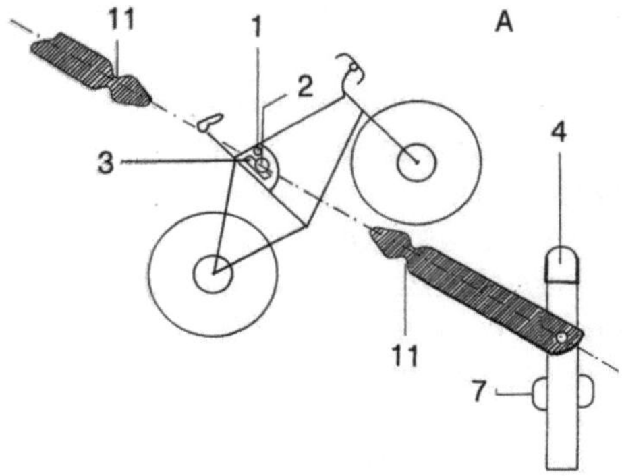

Fig. 9.25 Bicycle stand and security system (DE 20 2017 000 081 U1)

9.19.24 Example: Bicycle Stand and Security System

The utility model specification DE 20 2017 000 081 U1 (Title: New Bicycle Stand and Security System) describes the task of providing a security system for a bicycle to prevent theft (see Fig. 9.25). (DPMA, https://depatisnet.dpma.de/DepatisNet/depatisnet?action=pdf&docid=DE202017000081U1&xxxfull=1, accessed on 24 April 2026)

The main claim reads:

1. ***New*** [Point 1] *bicycle stand and security system, characterized by* ***division into two parts of previous security systems, (cables, chains, folding locks)*** [Point 2]. [Point 3] *One part is permanently connected to* ***the*** *bicycle* [Point 4] . *One part is permanently installed in* ***the*** *stand installation* [Point 5] . ***Drawing A (2) (4) (11)*** [Point 6]. *The* ***interfitting parts (2) and (11)*** [Point 7] *provide the stand and security for the bicycle.* ***Depending on the stand installation, pin (11) can be permanently attached to the bicycle or the stand installation. The same applies to the pin receptacle (2), depending on the desired design of the stand installation*** [Point 8]. *(DPMA, https://depatisnet.dpma.de/DepatisNet/depatisnet?action=pdf&docid=DE202017000081U1&xxxfull=1, accessed on 24 April 2026)*

Point 1: "new": The novelty of the subject matter of the main claim is a prerequisite for the legal validity of the utility model. It is not surprising that the applicant assumes novelty. However, a claim does not become novel simply because the applicant designates it as such. The novelty of a utility model is ultimately determined in a utility model cancellation procedure.[51] There is no official examination procedure for novelty and inventive step for a utility model.[52] However, an official search for relevant prior art documents can be requested.[53]

[51] Section 15 Utility Model Act.

[52] Section 8 (1) sentence 2 Utility Model Act.

[53] Section 7 (1) Utility Model Act.

Point 2: "Division into two parts of previous security systems, (cables, chains, folding locks)": A claim should not describe the prior art, nor how it can be further developed, but rather specify the features that realize the invention.

Point 3: A claim is a sentence, not several sentences. This is also intended to symbolize that a claim can protect only a single invention, not several.

Point 4: "the bicycle": A bicycle is introduced for the first time in the claim, so the indefinite article should be used. If a subject is introduced for the first time in a dependent claim that has not yet been introduced in any of the referenced claims, the indefinite article should also be used. If a term has already been introduced in a referenced claim, for example in the main claim, and is taken up again in a dependent claim, the definite article should be used.

Point 5: "the stand installation": The subject is introduced for the first time in the claim, so the indefinite article must be used.

Point 6: References to drawings or parts of the description such as "Fig. 1" or "see description for Fig. 1" or "see description on page 1" should be avoided.[54] This contradicts the principle that a claim should be understandable on its own.[55]

Point 7: "the interfitting parts (2) and (11)": The first question here is which parts are meant. Previously, "one" part and then again "one" part were described. However, it is not clear that these parts are the two parts that are then interfitted. It would have been possible to define a "first" part and a "second" part beforehand and then refer to these parts and describe that the first part is inserted into the second part. The next question is why the parts should be interfitted. Is this intended to enable a connection? In this case, it would make sense to describe a detachable connection of the first part with the second part. "Interfitting" does not necessarily result in a fixed connection. Moreover, this would significantly limit the scope of protection. Circumvention solutions could arise that cannot be countered by the main claim.

Point 8: A solution with a pin and a pin receptacle is described. This restricts the scope of protection of the main claim to such an extent that it appears almost worthless. A circumvention solution would be any kind of projection that can be inserted into a complementary receptacle and held by it. A possible variant that would not fall within the scope of protection would be, instead of a pin, a cylindrical or dovetail-shaped projection. The specific embodiment as a pin would be a typical feature for a dependent claim, since this specific embodiment offers particular advantages that allow a distinction from the prior art.

---

[54] Section 9 (8) Patent Ordinance.

[55] Section 14 sentence 1 Patent Act.

The corrected main claim reads:

1. Bicycle stand and security system, **characterized by:**

a first part, which is permanently connected to a bicycle, and
a second part, which is permanently connected to a stand installation, wherein the first part can be detachably connected to the second part in order to prevent theft of the bicycle and/or to provide a stable stand for the bicycle.

9.19.25 Example: Backpack for Carrying a Bicycle

Document DE 20 2017 004 477 U1 (Title: Backpack with Mounts for Carrying a Bicycle) describes the objective of carrying a bicycle with a backpack in such a way that the weight of the bicycle is evenly distributed over the carrier's body. This allows the carrier to transport the bicycle through rough terrain for a long period of time without significant impairment of balance (see Figs. 9.26 and 9.27, DPMA, https://depatisnet.dpma.de/DepatisNet/depatisnet?action=pdf&docid=DE202017004477U1&xxxfull=1, accessed on 24 April 2026).

The main claim of the utility model reads as follows:

1. Backpack with ***mounts for carrying a bicycle*** [Point 1], *which*

comprises a frame and
at least two shoulder straps,

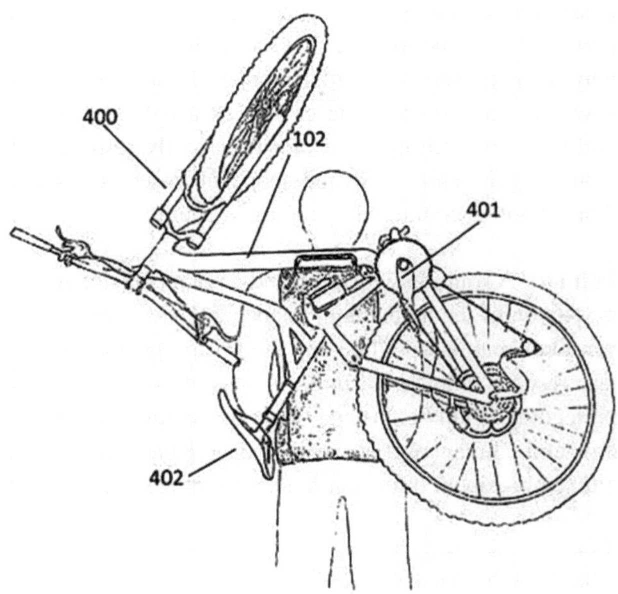

Fig. 9.26 Backpack for carrying a bicycle Fig. 4 (DE 20 2017 004 477 U1)

Fig. 9.27 Backpack for carrying a bicycle Fig. 6 (DE 20 2017 004 477 U1)

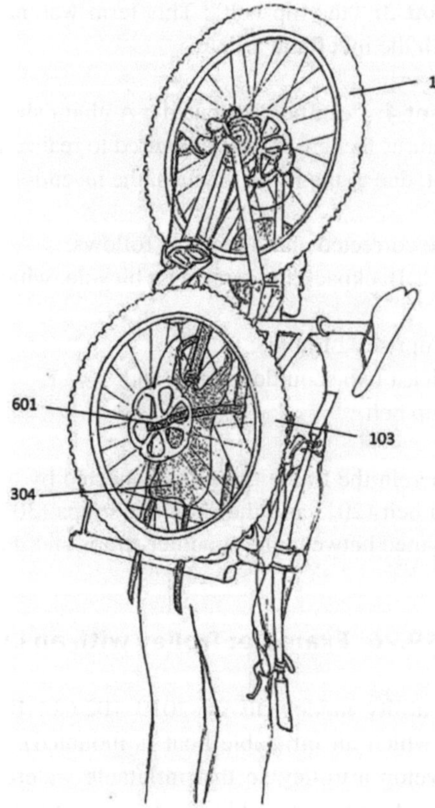

characterized in** **that [Point 2]

the frame (200) is connected by one or more flexible or rigid connections to **the hip belt** [Point 3] *(201) and the shoulder straps (301), so that the load of the bicycle (101) can be distributed between the shoulder straps and the hip belt (201)* ***can be distributed*** [Point 4]. *(DPMA, https://depatisnet.dpma.de/DepatisNet/depatisnet?action=pdf&docid=DE202017004477U1&xxxfull=1, accessed on 24 April 2026)*

Point 1: "Mounts for carrying a bicycle": "Mounts" are introduced, which are not required by the subsequent features. These mounts are therefore not necessary for the realization of the invention and can be omitted.

Point 2: "characterized in that": As a rule, the word "wherein" should be used instead of this formulation.[56]

[56] Section 9(1) sentence 1 of the Patent Ordinance.

Point 3: "the hip belt": This term was newly introduced and should therefore be used with the indefinite article.

Point 4: "can be distributed": A claim should not describe what can be, but what is. The features that are actually required to realize the invention must be specified. This expresses that, due to the listed features, the invention is not possibly, but regularly realized.[57]

The corrected claim reads as follows:

1. Backpack for carrying a bicycle, which

comprises a frame,
at least two shoulder straps, and
a hip belt,

wherein the frame (200) is connected by one or more flexible or rigid connections to the hip belt (201) and the shoulder straps (301), so that the load of the bicycle (101) is distributed between the shoulder straps and the hip belt (201).

9.19.26 Example: Trolley with an Inflatable Boat

In utility model DE 20 2019 102 622 U1 (Title: Trolley with integrated accessories, to which an inflatable boat is mounted), the technical problem addressed is to further develop a trolley so that inflatable watercraft, such as canoes or kayaks, can be transported with it. In addition, rapid assembly and disassembly of the inflatable watercraft should be possible.

The essential core of the invention is the attachment of an inflatable watercraft to a trolley. The invention is largely anticipated by the problem statement, which gives the impression that the invention lacks an inventive step. The invention is suggested by the problem statement, and the utility model therefore does not appear to be legally valid.[58] The description of the problem should be limited to stating that a "time-efficient use of inflatable watercraft is to be achieved," in order to leave sufficient room for an inventive step by the invention (see Fig. 9.28, DPMA, https://depatisnet.dpma.de/DepatisNet/depatisnet?action=pdf&docid=DE202019102622U1&xxxfull=1, accessed on 24 April 2026).

The set of claims 1 to 4 reads as follows:

1. Trolley ***with integrated accessories*** [Point 1], ***to which an inflatable boat is mounted*** [Point 2], *characterized in that* ***it*** [Point 3] *is connected to the* ***boat*** [Point 6] *by means of magnetic surfaces, hook-and-loop fasteners* ***or*** [Point 4] ***telescopic rods*** [Point 5].

[57] Section 34(4) Patent Act or Article 83 EPC.

[58] Section 1(1) Utility Model Act.

Fig. 2

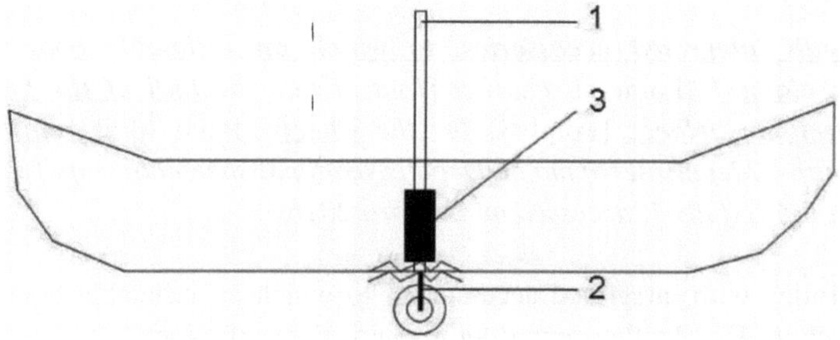

Fig. 9.28 Trolley with an inflatable boat (DE 20 2019 102 622 U1)

(DPMA, https://depatisnet.dpma.de/DepatisNet/depatisnet?action=pdf&docid=DE202019102622U1&xxxfull=1, accessed on 24 April 2026)

Point 1: "with integrated accessories": What are the integrated accessories? This is an incomprehensible feature that does not provide any positive effect for the applicant, but at most reduces the scope of protection.

Point 2: "to which an inflatable boat is mounted": The feature that a trolley is connected to an inflatable boat can already be considered the invention. However, the applicant has placed this feature in the preamble, which means that, in the applicant's view, it belongs to the prior art, i.e., it was already known. The use of the "characterized in that" formulation can therefore create a disadvantageous impression for the applicant of a lack of inventive step. It is advisable not to use the "characterized in that" formulation, but rather the word "wherein," which does not suggest a division between the prior art and the invention.

Point 3: "it": "It" most likely refers to the trolley. To avoid ambiguities, it is advisable to write out the terms in full.

Point 4: "or": Due to the use of "or", for example, a trolley with a magnet together with a hook-and-loop fastener is excluded from the scope of protection. However, this variant already represents a viable workaround. Instead of "or", "and/or" is better.

Point 5: "Telescopic rods": Unclear and therefore not feasible. In what way are telescopic rods used? After all, telescopic rods do not automatically create a connection between the inflatable boat and the trolley. The applicant probably intended a combination of magnet or hook-and-loop fastener and telescopic rods. However, such an embodiment is excluded by the use of "or" (see Point 4).

Point 6: "Boat": It is advisable to use the word "inflatable boat" instead of "boat" again, in order to avoid ambiguities from the outset.

*2. **Trolley with integrated accessories, to which an inflatable boat is mounted, according to claim 1*** [Point 1], *characterized in that the **hull of the boat*** [Point 2] *at the **connection surfaces*** [Point 3] *for the **wheels*** [Point 4] ***is reinforced*** [Point 5]. *(DPMA, https://depatisnet.dpma.de/DepatisNet/depatisnet?action=pdf&docid=DE202019102622U1&xxxfull=1, accessed on 24 April 2026)*

Point 1: "Trolley with integrated accessories, to which an inflatable boat is mounted, according to claim 1": "Trolley according to claim 1" is sufficient.

Point 2: "Hull of the boat": What is meant by the "hull" of a boat?

Point 3: "Connection surfaces": What are the connection surfaces?

Point 4: "the wheels": Wheels have not yet been introduced. For this reason, the definite article "the" cannot be used here. Furthermore, it is unclear which wheels are meant. They could be the wheels of the trolley, but this does not necessarily have to be the case. Unfortunately, the claim leaves much room for speculation, which is disadvantageous. Care should be taken to ensure a clear and precise description. In claims, every word must be carefully considered.

Point 5: "reinforced": It is unclear in what way something is to be reinforced. Therefore, this is a task-oriented formulation.

*3. Trolley with integrated accessories, to which an inflatable boat is mounted according to **claim 2*** [Point 1], *characterized in that the magnetic surfaces are **corresponding surfaces*** [Point 2] *on the **chassis*** [Point 3] *are congruent. (DPMA, https://depatisnet.dpma.de/DepatisNet/depatisnet?action=pdf&docid=DE202019102622U1&xxxfull=1, accessed on 24 April 2026)*

Point 1: It is not clear why claim 3 does not refer back to claims 1 and 2. A formulation such as "Trolley according to one of claims 1 or 2" would be appropriate. With this formulation "according to one of claims 1 or 2", the features of claim 3 can be combined with those of claim 1 or those of claim 2 or with those of claims 1 and 2 to form a new main claim if it is necessary to distinguish the invention from the prior art during the patent grant procedure.

Point 2: "corresponding surfaces": What are the corresponding surfaces or where are these corresponding surfaces?

Point 3: "Chassis": What is the chassis? The chassis could be a part of the trolley or the inflatable boat or some other device on another object.

4. Trolley with integrated accessories, to which an inflatable boat is mounted, ***according to claims 1 and 2*** [Point 1], *characterized in that* ***the chassis*** [Point 2] *is moved on two* ***free-running wheels*** [Point 3]. *(DPMA, https://depatisnet.dpma.de/DepatisNet/depatisnet?action=pdf&docid=DE202019102622U1&xxxfull=1, accessed on 24 April 2026)*

Point 1: "according to claims 1 and 2": This formulation means that the features of claim 4 can only be combined with the features of claim 1 together with the features of claim 2 to form a new main claim. It is therefore not possible to combine the features of claim 4 only with those of the main claim. The features of claim 2 must also necessarily be included. A combination of the features of claim 4 with those of claim 3 is also excluded. With regard to the reference, it is relatively easy to always use the formulation "according to one of the preceding claims". In individual cases, this may be incorrect if, for example, an item is used that is only introduced in claim 3. Such an error can be corrected at any time in a patent grant procedure or in litigation. However, it is not possible to extend a reference to claims to which the relevant claim did not originally refer.

Point 2: "the chassis": The chassis was only introduced in claim 3. A reference to claims 1 and 2 is therefore incorrect. However, this is a deficiency that can be easily corrected in an official procedure and is therefore inconsequential.

Point 3: "free-running wheels": It is unclear what "free-running" wheels are. Are these wheels without a brake, or wheels whose brake is released, or is something completely different meant?

The corrected main claim reads:

1. Trolley to which an inflatable boat can be mounted, wherein the trolley can be connected to the inflatable boat by means of magnetic surfaces, hook-and-loop structures and/or telescopic rods.

By using the terms "can be mounted" and "can be connected", the scope of protection is broadened, since a suitably designed trolley is already claimed and not just a trolley to which an inflatable boat is currently attached.

Claims 2 to 4 are unclear. Therefore, a corrected version cannot be presented.

9.19.27 Example: E-Steg Boat Trailer

In document DE 20 2015 008 800 U1 (Title: E-Steg Boat Trailer), the technical objective addressed is to provide an electrically powered watercraft that additionally enables trans-

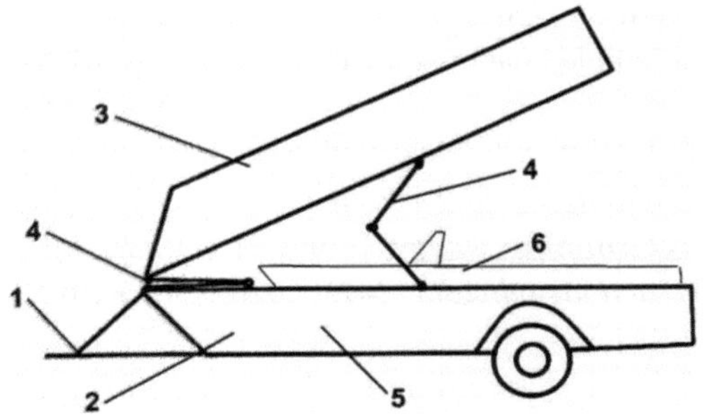

Fig. 9.29 E-Steg boat trailer (DE 20 2015 008 800 U1)

port over land (see Fig. 9.29, DPMA, https://depatisnet.dpma.de/DepatisNet/depatisnet?action=pdf&docid=DE202015008800U1&xxxfull=1, accessed on 24 April 2026).

The main claim of the utility model reads as follows:

1. ***Combination of boat trailer and floating dock for electrically powered watercraft/water sports equipment*** [Point 1], *characterized in that the* ***motor vehicle boat trailer*** [Point 2] ***serves both for road transport, and as a floating docking station*** [Point 3]. *(DPMA, https://depatisnet.dpma.de/DepatisNet/depatisnet?action=pdf&docid=DE202015008800U1&xxxfull=1, accessed on 24 April 2026)*

Point 1: "Combination of boat trailer and floating dock for electrically powered watercraft/water sports equipment": The designation already describes the inventive concept, namely the combination of a boat trailer with a docking station. By using the "characterized in that" formulation, the main claim is divided into a preamble, which the applicant considers to be prior art, and a characterizing portion. The designation of the subject matter, which already reveals the inventive concept, is in the preamble. As a result, the actual inventive concept is declared as already known, which creates a negative impression.

The fact that the combination of boat trailer and docking station is intended for an electrically powered watercraft appears arbitrary, since why should the subject matter of the claim not also be suitable for watercraft with other types of propulsion? This unnecessarily restricts the scope of protection considerably.

Point 2: "Motor vehicle boat trailer": So far, only a boat trailer has been described. For the same subject matter, at least within the set of claims, the same term should be used.

Point 3: "serves both for road transport and as a floating docking station": This functional feature has already been used in the designation of the subject matter of the claim. It is an unnecessary repetition. However, it would be necessary to describe which features the boat trailer possesses in order to be usable as a floating docking station. The claim is not executable and is therefore not protectable or legally valid.

The corrected main claim reads as follows:

1. Boat trailer for coupling to a vehicle, in particular a passenger car or truck, wherein the boat trailer is usable as a floating docking station.

9.19.28 Example: Automatic Emergency Reporting System

In this utility model DE 20 2015 005 961 U1 (Title: Automatic Emergency Reporting System with Accident Detection and Unconsciousness Detection), the technical objective is to reliably detect an accident involving a motorcyclist and his motorcycle and to trigger an alarm. (DPMA, https://depatisnet.dpma.de/DepatisNet/depatisnet?action=pdf&docid=DE202015005961U1&xxxfull=1, accessed on 24 April 2026)

Claims 1 to 3 of DE 20 2015 005 961 U1 read as follows:

1. Automatic emergency reporting system with accident detection and unconsciousness detection, ***which detects an accident by means of one or more sensors*** [Point 1], *characterized in that* ***one or more sensors*** [Point 2] *and* ***the radio system*** [Point 3] *are* ***connected to another system*** [Point 4] ***which is*** [Point 5], ***which determines the unconsciousness of the injured person*** [Point 6], *in order to verify the detected accident. (DPMA, https://depatisnet.dpma.de/DepatisNet/depatisnet?action=pdf&docid=DE202015005961U1&xxxfull=1, accessed on 24 April 2026)*

Point 1: "which detects an accident by means of one or more sensors": It is self-evident that an accident is detected by a sensor. This sensor is unfortunately not described in the characterizing portion. It is only stated that the sensor "is connected to a system". Overall, this feature can be omitted.

Point 2: "one or more sensors": If the sensors are meant from the preamble, the definite article should be used. It is not specified what is to happen with the sensors or how they function. The feature can therefore be omitted.

Point 3: "the radio system": So far, no radio system has been introduced. For this reason, the radio system must be used with the indefinite article.

Point 4: "another system": A new system is introduced without specifying how it is constructed. This is a functional feature, which is why the claim is not executable.

Point 5: "connected": The sensors, the radio system, and the "system" are connected. Why these elements of the emergency reporting system are connected and what may result from this remains unclear.

Point 6: "which determines the unconsciousness of the injured person": It seems to be the essential element of the invention that an alarm is only triggered after unconsciousness has been detected. However, it is not described how unconsciousness is determined. Thus, it remains at the statement of the objective, namely to provide a system that reliably determines the unconsciousness of the motorcyclist. The claim is not executable.

2. System according to claim 1, characterized in that at least one, several, or a combination of the following technologies is used for unconsciousness detection: sensors, mechanical devices, optical systems. (DPMA, https://depatisnet.dpma.de/DepatisNet/depatisnet?action=pdf&docid=DE202015005961U1&xxxfull=1, accessed on 24 April 2026)

Claim 2 is functional, as only possible sensors are specified without determining how these are to be used to detect unconsciousness.

3. System according to claim 1 or 2, characterized in that head movement is detected for unconsciousness detection. (DPMA, https://depatisnet.dpma.de/DepatisNet/depatisnet?action=pdf&docid=DE202015005961U1&xxxfull=1, accessed on 24 April 2026)

Claim 3 is not reproducible for the skilled person, since how is unconsciousness of the motorcyclist to be inferred from a head movement or the absence of a head movement?

9.19.29 Example: Sliding Sports Equipment

The utility model publication DE 20 2011 106 216 U1 (Title: Sliding sports equipment, skis, runners, vehicles or the like) addresses the task of optimizing and improving the gliding properties of skis or runners (see Figs. 9.30, 9.31, and 9.32, DPMA, https://depatisnet.dpma.de/DepatisNet/depatisnet?action=pdf&docid=DE202011106216U1&xxxfull=1, accessed on 24 April 2026).

In this utility model, formulations such as "or the like," "or similar type," "of all types," and "and the like" are frequently used. These formulations are of little help, as they do not provide the skilled person with concrete information for implementing the invention. They also do not lead to an extension of the scope of protection, since unclear statements do not allow for the determination of a scope of protection.

The main claim reads as follows:

1. Sliding sports equipment ***1a*** [Item 1], *ski 1a, skis 1a, or runners 1a, for locomotion and vehicles 1b* ***or the like*** [Item 2], *which on their underside 2 and/or top side 2 and/or outer surface 2, partially and/or over the entire surface, have a* ***sandfish skin surface profile*** [Item 3] *3a and/or a negative sandfish skin surface profile 3b, whereby these* ***increase the gliding ability by reducing the resistance of their contact surface, on snow surfaces, on ice surfaces, on land surfaces, on sand surfaces or water surfaces or the like and favorably influence these and/or thereby achieve an increase in speed and stabilization and/or a reduction in noise, when applied*** [Item 4]. *(DPMA, https://depatisnet.dpma.de/DepatisNet/depatisnet?action=pdf&docid=DE202011106216U1&xxxfull=1, accessed on 24 April 2026)*

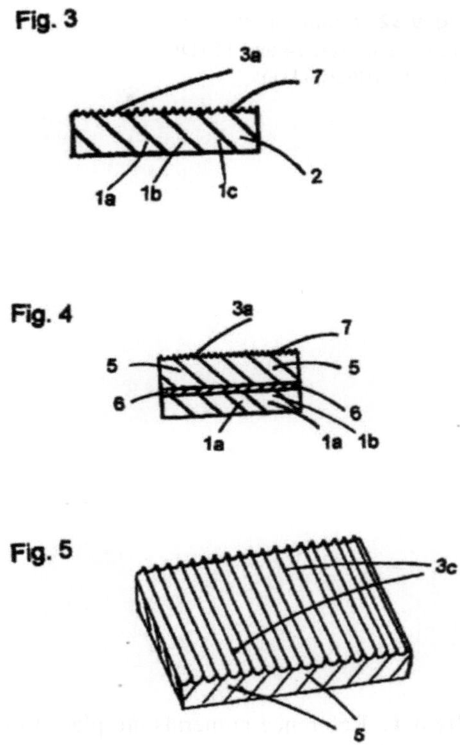

Fig. 9.30 Sliding sports equipment Figs. 3, 4, and 5 (DE 20 2011 106 216 U1)

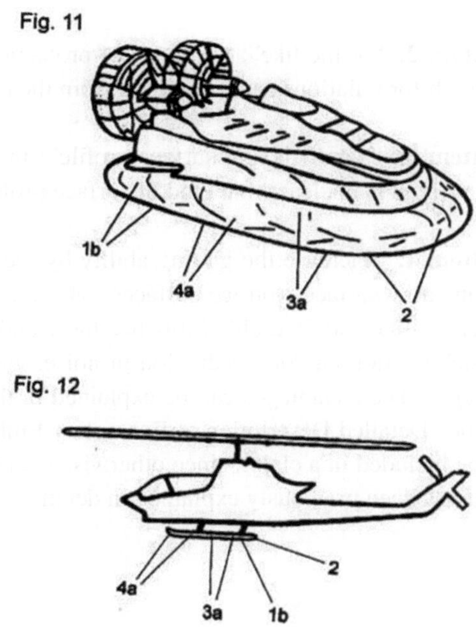

Fig. 9.31 Sliding sports equipment Figs. 11 and 12 (DE 20 2011 106 216 U1)

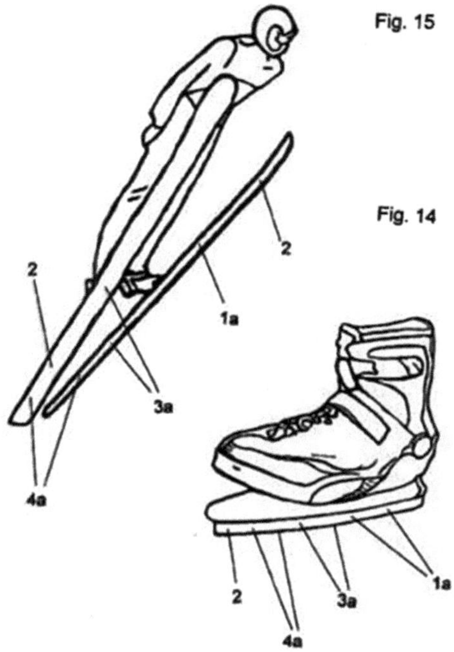

Fig. 9.32 Sliding sports equipment Figs. 14 and 15 (DE 20 2011 106 216 U1)

Item 1: Reference numerals are placed in parentheses in claims and not in parentheses in the "brief description of the drawings" and in the "detailed description of exemplary embodiments."

Item 2: "or the like": No scope of protection can be defined with indefinite statements. Such formulations can be omitted from the entire application documents.

Item 3: "Sandfish skin surface profile": In a claim, the features must be described precisely. It is unclear what kind of surface profile a "sandfish skin" has.

Item 4: "increase the gliding ability by reducing the resistance of their contact surface, on snow surfaces, on ice surfaces, on land surfaces, on sand surfaces or water surfaces or the like and favorably influence these and/or thereby achieve an increase in speed and stabilization and/or a reduction in noise, when applied": This passage states the advantages. The advantages can be explained in the section "Summary of the Invention" or in the "Detailed Description of Exemplary Embodiments". As a rule, advantages should not be included in a claim, since otherwise the claim becomes a statement of purpose, unless it has been previously explained in detail how the advantage is achieved.

9.19.30 Example: Emergency Buoy

The utility model specification DE 20 2011 002 521 U1 (Title: Emergency Buoy) addresses the task of providing a container for emergencies that is compact and can be carried relatively comfortably by a person either on the back or in the hand. The invention is intended to enable emergency care for injured persons who are in spatially confined, difficult situations (see Figs. 9.33 and 9.34, DPMA, https://depatisnet.dpma.de/DepatisNet/depatisnet?action=pdf&docid=DE202011002521U1&xxxfull=1, accessed on 24 April 2026).

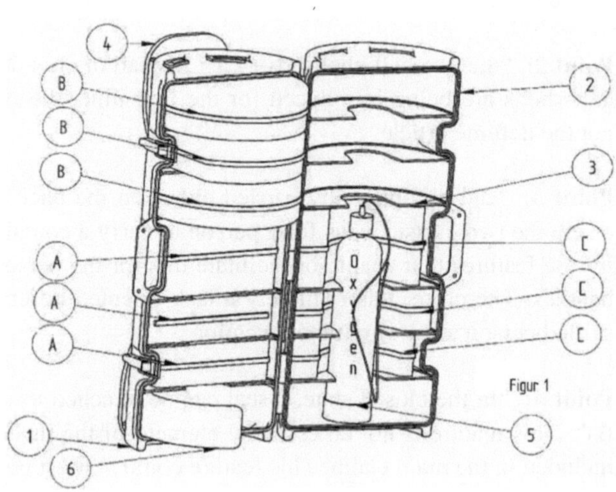

Fig. 9.33 Emergency buoy Fig. 1 (DE 20 2011 002 521 U1)

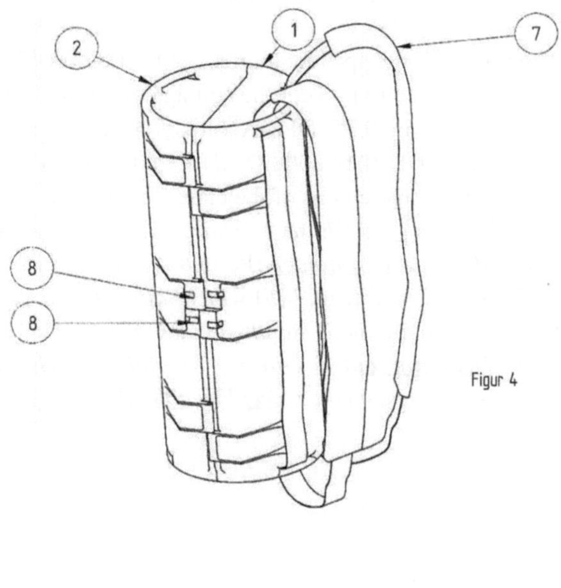

Fig. 9.34 Emergency buoy Fig. 4 (DE 20 2011 002 521 U1)

The main claim reads:

1. Sealable buoy for holding first aid supplies and medications for the initial treatment of patients and accident victims. [Point 1] *Characterized in that* **the two half-shells (1. + 2.)** [Point 2] *form a cylinder in the closed state by means of a hinge and can be optionally carried* ***either on the back or in the hand*** [Point 3]. ***In the closed state, a seal can be attached through a hole in the handle (detail B)*** [Point 4]. *(DPMA, https://depatisnet.dpma.de/DepatisNet/depatisnet?action=pdf&docid=DE202011002521U1&xxxfull=1, accessed on 25 April 2026)*

Point 1: A claim consists of only one single sentence. An invention must be described in one sentence. This also expresses that a claim conveys only one inventive concept.

Point 2: "the two half-shells (1. + 2.)": Instead of (1. + 2.), it is better to use (1, 2). Since half-shells are being introduced for the first time, the indefinite article should be used, not the definite article.

Point 3: "can be optionally carried either on the back or in the hand": These are precisely the two classic ways for a person to carry a container. It would be correct to specify the features that enable or facilitate this for the person, for example, special carrying handles. Therefore, this is merely a task-oriented description, which does not enable a skilled person to carry out the invention.

Point 4: "In the closed state, a seal can be attached through a hole in the handle (detail B)": This feature is not an essential element of the invention. It should by no means be included in the main claim. This feature could at most be included in a dependent claim.

9.19.31 Example: Drinking Aid

The utility model specification DE 20 2008 000 979 U1 (Title: Drinking Aid) describes a device intended to prevent the spillage of a liquid from a cup, for example on an unstable surface (see Fig. 9.35, DPMA, https://depatisnet.dpma.de/DepatisNet/depatisnet?action=pdf&docid=DE202008000979U1&xxxfull=1, accessed on 25 April 2026).

The first three claims of the claim set are as follows:

1. ***Drinking aid*** [Point 1], *characterized in that a drinking cup* ***(Fig. 2–6)*** [Point 2] *is suspended in a* ***cup holder*** [Point 1] *(Fig. 1–1) via two opposing pivot points (Fig. 1–2) and a swivel ring (Fig. 1–5) which, offset by 90 degrees, has two opposing pivot points (Fig. 1–3), in a gimbal-mounted manner.*

2. ***Drinking aid*** *according to claim 1, characterized in that the drinking cup (Fig. 2–6) is designed as a* ***replaceable, thermally insulated vessel*** [Point 3].

Fig. 9.35 Drinking aid Fig. 1 and 2 (DE 20 2008 000 979 U1)

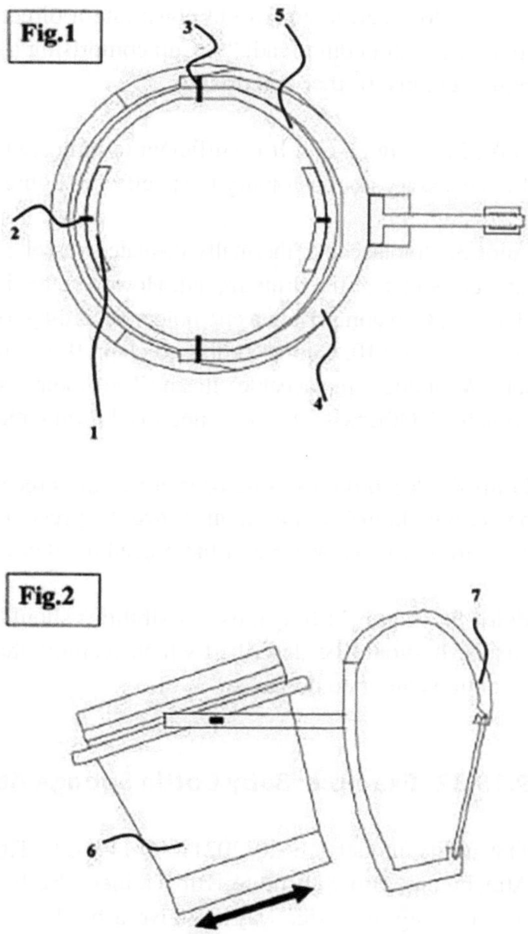

3. *Drinking aid* ***according to one or more of the preceding claims,*** [Point 4] *characterized in that the outer claw (Fig. 1–4) is provided with a handle, which may be designed in the form of a handle or as a loop (Fig. 2–7)* ***can be*** [Point 5]. *(DPMA, https://depatisnet.dpma.de/DepatisNet/depatisnet?action=pdf&docid=DE20 2008000979U1&xxxfull=1, accessed on 25 April 2026)*

Point 1: A drinking aid is claimed. On the other hand, a drinking cup and a cup holder are described. It might be advisable not to introduce an additional term such as cup holder, but rather to claim this instead of the drinking aid. The term drinking aid could then be omitted. In this case, there would be a clear system of a cup holder and a drinking cup, with the cup holder being able to accommodate the drinking cup. It would also be sensible to claim the cup together with the cup holder in a further claim. In this way, a cup holder and a cup would constitute a direct potential infringement form. A corresponding claim could read: “9. Cup comprising a cup holder according to one of the preceding claims, further comprising ….”

Point 2: “(Fig. 2–6)”: It is sufficient to indicate the reference numeral, i.e., (6). It is neither necessary nor customary to specify the figure number.

Point 3: “replaceable, thermally insulated vessel”: This gives the impression that the drinking cup is part of the drinking aid. However, this is not the case. It would be better, after a claim “9. Cup comprising a cup holder according to one of the preceding claims …”, to provide a claim: “10. Cup according to claim 9, wherein the cup is a thermally insulated vessel.” What does “replaceable” mean? The features should be described so that replaceability is enabled. Otherwise, this is a functional formulation, which does not belong in a claim.

Point 4: “according to one or more of the preceding claims”: “according to one of the preceding claims” is sufficient. Since the preceding claims also contain back-references, “according to one or more of the preceding claims” is already included.

Point 5: “can be”: In claims, possibilities should not be indicated, but rather what actually is. It should be described which features the invention has, not which features the invention can or could have.

9.19.32 Example: Baby Bottle Sponge Attachment

The utility model DE 20 2021 000 115 U1 (Title: Drinking Aid, Baby Bottle Sponge Attachment, Bib Substitute, Bib Holder, Pacifier Aid) sets out the technical objective of providing an easier way to “give a bottle” to babies. In particular, an alternative to a bib is provided. The problem that a bib is often pulled off by babies is also solved in this way. The drinking aid comprises a sponge that lies around the infant’s mouth while drinking (see Fig. 9.36, DPMA, https://depatisnet.dpma.de/DepatisNet/depatisnet?action=pdf&docid=DE202021000115U1&xxxfull=1, accessed on 25 April 2026).

Reference numeral list

1. Nipple of the baby bottle
2. Baby bottle
3. Rail holder
4. Mushroom head for sponge attachment
5. Nubs of the adjustment aid
6. Sponge attachment ergonomically shaped towards the baby
7. Rail guide

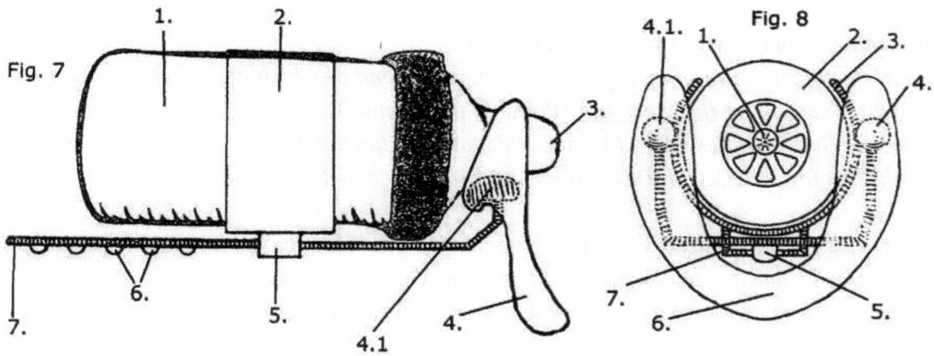

Fig. 9.36 Baby bottle sponge attachment (DE 20 2021 000 115 U1)

The designation of the reference numerals applies to Figure 8. In Figure 7, the reference numerals were labeled differently. This appears to be an error on the part of the applicant.

The claim set of the utility model comprises nine claims:

1. ***The*** [Point 1] ***combination*** [Point 2] *of an adjustable rail, at the end of which a mounted sponge is attached,* **to serve as a drip guard for baby bottles** [Point 3]. *(DPMA, https://depatisnet.dpma.de/DepatisNet/depatisnet?action=pdf&docid=DE20 2021000115U1&xxxfull=1, accessed on 25 April 2026)*

Point 1: "The": A combination is newly introduced in the claim set. For this reason, an indefinite article ("a") should be used instead of the definite article ("the"). However, in the designation of the subject matter of the claim, the use of an article is generally omitted.

Point 2: "Combination": A combination comprises two or more elements. The claim describes only a rail, i.e., only one item. Therefore, the word "combination" is used incorrectly.

Point 3: "to serve as a drip guard for baby bottles": The claim is unclear. It is not understandable how the rail is arranged on the baby bottle and how the rail is supposed to function as a drip guard. Possibly, the sponge is intended to be used as a drip guard. However, this cannot be inferred from the claim. It is also unclear in what way the rail is adjustable and why the adjustability is an essential feature of the invention that must be included in the main claim.

An alternative claim wording would be:

1. Drip guard for a baby bottle (2), comprising:

a rail holder (3) for at least partially gripping the baby bottle (2)
a rail (7) attached to the rail holder (3), and
a sponge (6) arranged at one end of the rail (7),

wherein the sponge (6) is arranged near the teat (1) so that the sponge (6) acts as a drip guard when a baby drinks from the baby bottle (2).

2. The combination of an adjustable rail, at the end of which ***commercially available*** [Point 1] *bib can be* ***inserted into*** [Point 2] ***the*** [Point 3] *ring eyelets to serve as a drip guard for baby bottles. (DPMA, https://depatisnet.dpma.de/DepatisNet/depatisnet?action=pdf&docid=DE202021000115U1&xxxfull=1, accessed on 25 April 2026)*

Point 1: "commercially available": The use of the feature "commercially available" unnecessarily narrows the scope of protection. Using a non-commercially available bib would be an easy way for an imitator to circumvent the protection right.

Point 2: "the": Ring eyelets have not yet been introduced in the set of claims. Therefore, the indefinite article should be used.

Point 3: "Ring eyelets": It is unclear where the ring eyelets are arranged and how they are to be used.

The claim does not refer back to the main claim. All nine claims lack any back-reference. The set of claims therefore comprises nine independent claims. For this reason alone, the set of claims would have no chance of surviving in a cancellation procedure against the utility model. Only one independent claim per claim category (device, method, and use) may be included in a set of claims. The remaining claims must be formulated with a back-reference to the independent claims and thus as dependent claims. On the other hand, suitable back-references could easily be formulated here, which would give the set of claims a meaningful internal structure.

An alternative claim wording would be:

2. Drip guard according to claim 1, wherein a bib for a baby is arranged on the rail.

Attaching the bib to the rail with ring eyelets is only one solution. Another possible attachment would be a hook-and-loop fastener. The possibilities for attaching the sponge and the bib can be described in the description or in a further dependent claim.

3. The ***curved*** [Point 1] *(ergonomically curved forward—facing the child)* ***sponge attachment*** [Point 2] *(**conical or. convex depending on the viewing direction*** [Point 3]*), which is modeled after the lower facial shape in the mouth area—and is skin-neutral (skin-friendly), highly absorbent, and washable—ensures* ***due to its shape*** [Point 4], *that when the baby bottle is applied, the mouth is enclosed laterally and below, so that no milk can run into the baby's neck fold. (DPMA, https://depatisnet.dpma.de/DepatisNet/depatisnet?action=pdf&docid=DE202021000115U1&xxxfull=1, accessed on 25 April 2026)*

Point 1: "curved": It would make sense to describe the overall shape of the sponge and then additionally explain how this particular bulge of the sponge is formed.

Point 2: "sponge attachment": In claim 1, a sponge 6 was introduced. "Sponge attachment" is a newly introduced term. It can be assumed that both terms refer to the same thing. Such ambiguities must be avoided at all costs. They always work to the detriment of the applicant.

Point 3: "(conical or convex depending on the viewing direction)": Conical means cone-shaped. Convex means curved outward. Depending on the viewing direction, convex could be interpreted as curved outward or inward. These designations make little sense. Possibly, it was meant that the sponge can be shaped in a curved manner, for example, to better catch liquid from the baby bottle. However, this feature is extremely poorly described, making the claim incomprehensible. Furthermore, in claims, apart from reference signs, nothing should be put in parentheses. If something is put in parentheses, it signals that it is not important. However, a claim should contain only important features. The use of parentheses in a claim is therefore to be rejected.

Point 4: "due to its shape": Due to which shape? Possibly because of a shape of the sponge that is "modeled after the lower facial shape in the mouth area." However, this is not clear; it could also be due to another shape.

An alternative claim wording would be:

3. Drip guard according to one of claims 1 or 2, wherein the sponge (6) is at least partially shaped complementary to the lower half of the baby's face and/or wherein the sponge (6) is highly absorbent and/or skin-neutral and/or washable.

The features "highly absorbent", "skin-neutral", and "washable" are probably not valuable fallback positions. These features can therefore alternatively be omitted.

4. The ***as described in*** [Point 1] ***Fig. 3 item. 2*** [Point 2] ***"nubs" of the adjustment aid*** [Point 3], ***which allow the rail to be adjusted with just one finger— of the hand currently holding the bottle— possible*** [Point 4]. ***By selecting a different "hole"— by moving the nubs of the adjustment aid into the rail— an extension of the rail is possible, so that larger bottles can also be used with the drinking aid*** [Point 5]. *(DPMA, https://depatisnet.dpma.de/DepatisNet/depatisnet?action=pdf&docid=DE202021000115U1&xxxfull=1, accessed on 25 April 2026)*

Point 1: "as in": So "similar to". In a claim, an exact, not a similar, description of the invention is required.

Point 2: "as described in Fig. 3 item 2": A reference to the drawings should be avoided as much as possible.

Point 3: "'Nubs' of the adjustment aid": A new term is introduced here, namely that of the adjustment aid. This is not necessary; the existing terms are sufficient, since the adjustment aid is actually an integral element of the rail. Adaptation to baby bottles of different lengths is indeed achieved by providing a way to shift the rail holder, which essentially serves as an attachment to the baby bottle, and the rail.

Point 4: "which allows the rail to be adjusted with just one finger—of the hand currently holding the bottle": This is a functional formulation. Such a formulation is not sufficient to define a feature. The structural properties of the features must be described, followed by an explanation of how these features interact. If the description of the structural features is omitted, the result is a purely functional formulation, which is generally not accepted as a clear and unambiguous description of a feature.

Point 5: "By selecting a different 'hole'—by moving the nubs of the adjustment aid into the rail—it is possible to extend the rail, so that larger bottles can also be used with the drinking aid": This particular embodiment can be included in the description to explain how the adjustability and latching are enabled. However, an inventive step cannot be established solely by the "nubs", as it is within the routine skill of the average practitioner to use such a latching mechanism.

An alternative claim formulation would be:

4. Drip guard according to one of the preceding claims, wherein the rail holder (3) is displaceable relative to the rail (7), so that the drip guard can be adapted to the length of the baby bottle (2).

5. The entire configuration of the drinking aid in its depicted form and function is to be protected. This includes the rail, the adjustment aid, as well as the ring loops for bibs or mushroom head mounts for a sponge. The function of the adjustment aid in combination with the adjustable rail is also worthy of protection. (DPMA, https://depatisnet.dpma.de/DepatisNet/depatisnet?action=pdf&docid=DE202021000115U1&xxxfull=1, accessed on 25 April 2026)

A claim must describe the inventive features. References to claimed objects do not belong in a claim. Claim 5 can be deleted.

6. The forked design of the adjustable rail (Fig. 2 and Fig. 2.1), as well as the accommodation of the sponge by mushroom heads (Fig. 5 + Fig. 6, point 5), or the accommodation for bibs by the ring loops, provide a stress-free bottle. (DPMA, https://depatisnet.dpma.de/DepatisNet/depatisnet?action=pdf&docid=DE202021000115U1&xxxfull=1, accessed on 25 April 2026)

Another interesting feature of the invention is the splitting of the rail at one end and the formation of thickenings at the endpoints of the split parts of the rail. However, this particular embodiment of the invention should be appropriately claimed.

An alternative claim formulation would be:

5. Drip guard according to one of the preceding claims, wherein the rail (7) has a thickening at one end to provide better support for the attached sponge (6) and/or wherein the rail (7) splits at this end and/or wherein the ends of the split sections have thickenings to provide better support for the attached sponge (6).

7. The ring loops for receiving commercially available bibs, which make it difficult or even impossible to tear off or pull away the bibs. (DPMA, https://depatisnet.dpma.de/DepatisNet/depatisnet?action=pdf&docid=DE202021000115U1&xxxfull=1, accessed on 25 April 2026)

Claim 7 describes the advantages, namely that tearing off or pulling away the bib is prevented. However, no features are described as to how the ring loops are to be arranged and how they interact with the rail.

8. The attachment of a drip guard in the depicted form as a whole is to be protected. (DPMA, https://depatisnet.dpma.de/DepatisNet/depatisnet?action=pdf&docid=DE202021000115U1&xxxfull=1, accessed on 25 April 2026)

In claim 8, a drip guard is claimed in a cursory manner. It cannot be said that the invention is described precisely and unambiguously in the claim so that a skilled person can carry it out. The claim is unclear and not legally valid.

9. The depicted design of all drawings presented in this application. (DPMA, https://depatisnet.dpma.de/DepatisNet/depatisnet?action=pdf&docid=DE202021000115U1&xxxfull=1, accessed on 25 April 2026)

Claim 9 is unsurpassed in its lack of clarity. "Designs of all drawings" are claimed. The claim can be deleted, as it is completely unclear. Furthermore, "designs" are claimed. Designs, i.e., aesthetic configurations, are not eligible for protection under patent law or utility model law.[59] Aesthetic designs can be protected by design law.

9.19.33 Example: Mobile Drinking Aid

Utility model DE 20 2017 002 000 U1 (Title: Mobile—Drinking Aid, double and single) sets out to provide a drinking aid so that persons with limited motor skills can drink without assistance (see Fig. 9.37, DPMA, https://depatisnet.dpma.de/DepatisNet/depatisnet?action=pdf&docid=DE202017002000U1&xxxfull=1, accessed on 25 April 2026).

The set of claims reads as follows:

*1. **Mobile drinking aid with one or two drinking bottles*** [Point 1]. ***The bottle baskets*** [Point 2] *are mounted on a **mobile and height-adjustable base*** [Point 3]. *(DPMA,* https://depatisnet.dpma.de/DepatisNet/depatisnet?action=pdf&docid=DE202017002000U1&xxxfull=1, *accessed on 25 April 2026)*

Point 1: "Mobile drinking aid with one or two drinking bottles": This sentence does not meet the requirements of a claim. No features are specified to realize the drinking aid. Instead, it is merely stated that the drinking aid should contain one or two drinking bottles. This means that simply placing two bottles side by side already constitutes a mobile

[59] Section 1, paragraph 3, no. 2 of the Patent Act or Section 1, paragraph 2, no. 2 of the Utility Model Act.

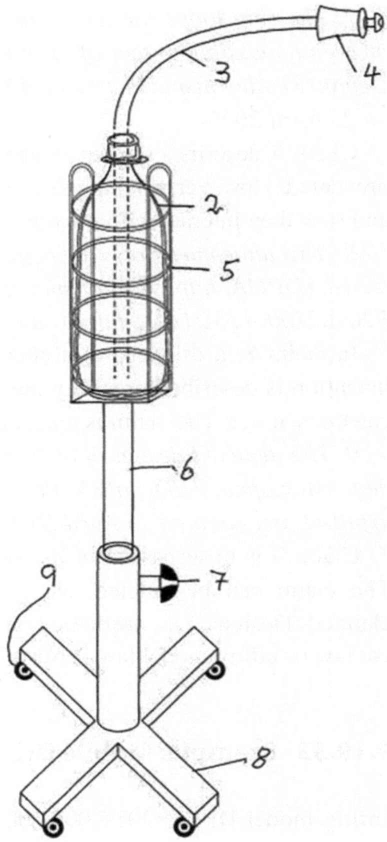

Fig. 9.37 Mobile drinking aid (DE 20 2017 002 000 U1)

drinking aid according to the main claim. Even a single drinking bottle qualifies as a mobile drinking bottle according to the wording, since both a single and two bottles can be moved and are therefore mobile. It is evident that the arrangement of two drinking bottles constitutes an infringement of utility model claim 1. However, it is equally obvious that the main claim cannot be legally valid. A claim is described by a sentence. The first sentence can therefore be regarded as the main claim. The second sentence can be interpreted as a dependent claim.

Point 2: "The bottle baskets": Bottle baskets are newly introduced here, so the indefinite article should be used. Furthermore, the question arises as to why several bottle baskets are described. If the drinking aid comprises only one bottle, only one bottle basket is needed. In addition, claim 2 states that two bottles can even be arranged in one bottle basket. The use of the plural thus appears even more puzzling in light of the technical teaching of claim 2.

Point 3: "mobile and height-adjustable base": What does the base have to do with the drinking aid? Is it a part of the drinking aid?

An alternative claim formulation would be:

1. Drinking aid for supplying a person with liquid from a bottle (2), comprising:

a basket (5) in which the bottle (2) can be arranged, and
a device on which the basket (5) is arranged, wherein the device is mounted on rollers so that the device is mobile and/or wherein the device is configured such that the basket (5) is height-adjustable.

2. Drinking aid according to claim 1, characterized in that one or two drinking bottles are held in ***a*** [Point 1] *bottle basket* [Point 2]. *(DPMA, https://depatisnet.dpma.de/DepatisNet/depatisnet?action=pdf&docid=DE202017002000U1&xxxfull=1, accessed on 25 April 2026)*

Point 1: In claim 1, the definite article was used for the bottle baskets. In claim 2, the indefinite article is used. This is the wrong way around. The indefinite article should be used when a term is introduced for the first time in a claim. If the already introduced term is used again, the definite article should be used.

Point 2: "one or two drinking bottles are held in **a** bottle basket": No features are specified as to how a bottle basket is designed so that it can alternatively accommodate one or two drinking bottles.

The features of claim 2 are already included in the main claim. Claim 2 can therefore be omitted.

3. Drinking aid according to claim 1 [Point 1], *characterized in that a* ***flexible*** [Point 2] ***drinking tube is inserted into the drinking bottle by means of a bottle closure*** [Point 3]. *(DPMA, https://depatisnet.dpma.de/DepatisNet/depatisnet?action=pdf&docid=DE202017002000U1&xxxfull=1, accessed on 25 April 2026)*

Point 1: All dependent claims of the claim set refer exclusively to the main claim ("… according to claim 1…"). As a result, in a cancellation procedure, only the respective dependent claim and the main claim can be combined to obtain a new main claim. It is then not permissible, for example, to combine claim 3, claim 2, and claim 1. It is therefore better to use references that allow any combination of the claims, in particular by using the wording "…according to any of the preceding claims…".

Point 2: "flexible": The use of a rigid drinking tube, which for example includes a bend or only a flexible section, would already be a possible workaround. The adjective "flexible" does not provide an increased inventive step for the claim. On the other hand, the scope of protection is significantly restricted. It is therefore not advisable to use this adjective.

Point 3: "drinking tube is inserted into the drinking bottle by means of a bottle closure": How is a drinking tube inserted by means of a bottle closure? Why is a so-called "bottle closure" provided, and does a "bottle closure" not result in the bottle no longer being drinkable from, even if a drinking tube is still present? What is the purpose of a bottle closure, since a bottle closure contradicts the use of a drinking tube? Such ambiguities should be avoided.

Alternative wording for claim 3:

2. Drinking aid according to claim 1, wherein a drinking tube (3) is arranged on the drinking bottle (2) in such a way that the person can drink from the bottle (2).

4. Drinking aid according to claim 1, characterized in that at the outlet of the flexible ***drinking tube*** [Point 1], *an* ***ergonomic*** [Point 2] *mouthpiece is attached. (DPMA, https://depatisnet.dpma.de/DepatisNet/depatisnet?action=pdf&docid=DE202017002000U1&xxxfull=1, accessed on 25 April 2026)*

Point 1: "drinking tube": A new term is introduced. Therefore, an indefinite article should be used. However, it can be assumed that the drinking tube refers to the drinking hose. Care should be taken to use terms consistently in the claim set and throughout the entire application.

Point 2: "ergonomic": It would be advisable to describe more precisely how the mouthpiece is shaped.

Alternative wording for claim 4:

3. Drinking aid according to one of claims 1 or 2, wherein a mouthpiece for sucking the liquid from the bottle (2) is arranged at the outlet of the drinking tube (3) and/or wherein the mouthpiece is ergonomically shaped.

Through the "and/or" connection, the mouthpiece can be included in the main claim without the feature of ergonomic shaping. However, with an "and/or" combination, nonsensical alternatives may arise. What is the ergonomic shaping of the second half-sentence without the mouthpiece for sucking the liquid of the first half-sentence? This issue can easily be resolved in an official or court procedure by omitting the nonsensical alternative. However, nothing can be added that was not previously included. It is therefore always advisable to use an "and/or" connection of two features rather than an "and" or "or" connection.

5. Drinking aid according to claim 1, characterized in that ***one or two*** [Point 1] ***metal mesh baskets*** [Point 2] *are arranged on a* ***base frame*** [Point 3]. *(DPMA, https://depatisnet.dpma.de/DepatisNet/depatisnet?action=pdf&docid=DE202017002000U1&xxxfull=1, accessed on 25 April 2026)*

Point 1: "one or two": It is sufficient to describe one metal mesh basket. With "one" or "a", two, three, four, or any number of metal mesh baskets are also included. However, if a specific number is to be presented as particularly inventive, then, for example, two metal baskets should actually be described. Here, however, it is not apparent that two and not three or four or five metal baskets are relevant. An arrangement with two metal baskets does not involve a higher inventive step than an arrangement with one or three metal baskets.

Point 2: "metal mesh baskets": Metal mesh baskets are introduced. The question arises as to whether this refers to the bottle basket of the main claim. It is therefore unclear whether an additional feature, namely that the basket is designed as a metal mesh, is to be described. It should always be borne in mind that consistency of terms within the claim set is a basic prerequisite for a valuable property right.

Point 3: "base frame": A new term "base frame" is introduced. The question arises as to whether and in what way the base frame is connected to the drinking aid. It is also questionable whether a base frame is actually a new feature, since a base was described in the main claim. Furthermore, the question arises as to why the metal mesh baskets are arranged on the base frame. What function does the base frame serve? Finally, the feature of the base frame in claim 5 appears superfluous, since it is already included in the main claim that the bottle baskets are mounted on a base.

An alternative wording for claim 5 would be:

4. Drinking aid according to any of the preceding claims, wherein the bottle basket is designed as a metal mesh.

6. Drinking aid according to claim 1, characterized in that the ***single or double metal mesh basket*** [Point 1] *is permanently mounted on a* ***two-part*** [Point 2] ***telescope*** [Point 3] . *The telescope is height-adjustable. (DPMA, https://depatisnet.dpma.de/DepatisNet/depatisnet?action=pdf&docid=DE202017002000U1&xxxfull=1, accessed on 25 April 2026)*

Point 1: "single or double metal mesh basket": What is a single metal mesh basket? What, then, is a complicated metal mesh basket? What is a double metal mesh basket? Which features must be designed in what way to result in a double metal mesh basket?

Point 2: "two-part": What is a two-part telescope? Does a two-part telescope simply consist of two parts, or is there more implied? Is it perhaps meant that the two-part design allows for height adjustability? Claim 6 only leads to speculation. For a claim to be enforceable before an infringement court, it must be clearly and unambiguously formulated. This necessary clarity is lacking in claim 6, so claim 6 is legally without significance.

Point 3: "Telescope": A new element is introduced again, similar to the base frame in claim 5. The question also arises as to how the telescope is related to the drinking aid.

An alternative wording for claim 6 would be:

5. Drinking aid according to one of the preceding claims, wherein the basket is telescopically height-adjustable.

9.19.34 Example: Hot Dog Drinking Cup

The utility model publication DE 20 2016 002 669 U1 (Title: Hot Dog (Curry) Sausage Croissants Drinking Cup) addresses the technical problem of being able to hold a drinking cup and a sausage or a croissant in just one hand. Advantageously, the free hand can be used to carry a shopping bag, a handbag, or a suitcase (see Fig. 9.38 and Table 9.1, DPMA, https://depatisnet.dpma.de/DepatisNet/depatisnet?action=pdf&docid=DE202016002669U1&xxxfull=1, accessed on 25 April 2026).

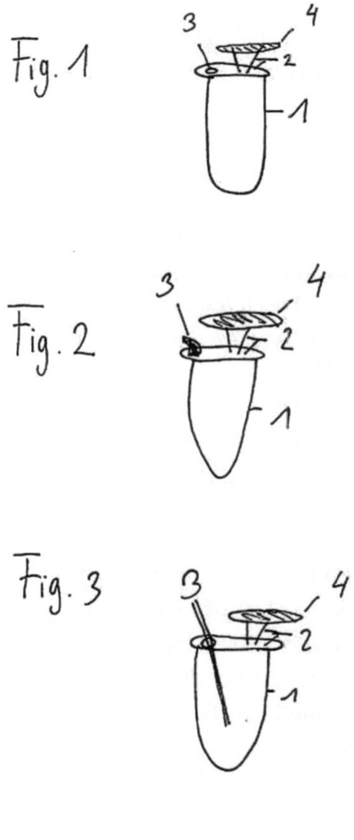

Fig. 9.38 Hot Dog Drinking Cup (DE 20 2016 002 669 U1)

Table 9.1 Hot Dog Drinking Cup—Reference Number List (DE 20 2016 002 669 U1)

Reference list	
1	Holder for sausage, hot dog, currywurst, or croissant
2	The attachment lid
3	Opening for e.g. drinking straw
4	Sausage or croissant
5	(Plastic, cardboard) drinking cup made of

The set of claims comprises three protection claims:

1. Drinking cup ***with including*** [Point 1] *attachment lid for holding* ***sausage, hot dogs, croissants*** [Point 2] ***characterized in that, that*** [Point 3] *the drinking cup with including attachment lid (2), which enables holding hot dogs, all kinds of sausages, croissants. (DPMA, https://depatisnet.dpma.de/DepatisNet/depatisnet?action=pdf&docid=DE202016002669U1&xxxfull=1, accessed on 25 April 2026)*

Point 1: "with including": "With" and "including" are synonyms. Therefore, it is sufficient to use the word "with" and omit the word "including".

Point 2: "Sausage, hot dogs, croissants": Infringement courts now only recognize infringement if there is a literal infringement. Therefore, a coffee cup lid on which a pretzel can be placed would not constitute an infringement of the main claim.

Point 3: "characterized in that": The two-part claim formulation separates a preamble from a characterizing part. The preamble represents the closest prior art, i.e., what was already known and from which the inventor started when creating the invention. The characterizing part describes the actual invention. Here, the preamble is: "Drinking cup with including attachment lid for holding sausage, hot dogs, croissants." The characterizing part is: "the drinking cup with including attachment lid (2), which enables holding hot dogs, all kinds of sausages, croissants". In substance, the preamble and the characterizing part are therefore identical, so the applicant is telling an infringement court or a patent office that the invention is already known, i.e., belongs to the prior art, which is, of course, not what the applicant intends.

An alternative wording of the main claim would be:

1. Attachment lid (2) for a drinking cup (1), comprising a holder for a food item (4), in particular a sausage, a hot dog, or a croissant.

By using a "in particular" formulation, examples can be included that can, if necessary, be used to distinguish from relevant prior art.

2. ***Attachment*** [Point 1] *according to claim 1, characterized in that* ***the drinking cup with including attachment lid (2) has an opening (3)*** [Point 2], *through which the liquid* *can be taken in via the mouth. (DPMA, https://depatisnet.dpma.de/DepatisNet/depatisnet ?action=pdf&docid=DE202016002669U1&xxxfull=1, accessed on 25 April 2026)*

Point 1: "Attachment": A new term is introduced. The question arises as to whether the attachment refers to the attachment lid. Such ambiguities should definitely be avoided. Terms should be used consistently in the set of claims and throughout the entire application.

Point 2: "the drinking cup with including attachment lid (2) has an opening (3)": Who actually has the opening? The drinking cup and the attachment lid, or only the attachment lid?

Claim 2 refers to a new subject matter, namely an attachment lid and not, as in claim 1, to a drinking cup. Claim 2 is therefore an independent claim.

An alternative wording for claim 2 could be:

2. Attachment lid (2) for a drinking cup according to claim 1, wherein the attachment lid (2) has an opening (3) for drinking.

3. Attachment lid ***according to one of the preceding claims*** [Point 1] *characterized in that the attachment lid (2) has an opening (3), through which a drinking straw* ***can pass through without difficulty*** [Point 2]. *(DPMA, https://depatisnet.dpma.de/DepatisNet/depatisnet?action=pdf&docid=DE202016002669U1&xxxfull=1, accessed on 25 April 2026)*

Point 1: Claim 3 cannot refer to the main claim, as the latter concerns a drinking cup. The reference in claim 3 must therefore read: "according to claim 2".

Point 2: What does "without difficulty" mean? Is some squeezing still allowed, or must the drinking straw be able to fall through the opening into the drinking cup without friction? At this point, it might be useful to specify diameter ranges for the opening. For example, it could be formulated: "Opening with a diameter between 5 mm and 20 mm, in particular with a diameter between 10 mm and 15 mm, more particularly with a diameter of 13 mm".

9.19.35 Example: Drinking Aid for Mobile and Stationary Use

Utility model DE 20208487 U1 (Title: Drinking aid for mobile and stationary use) describes the technical objective of providing a drinking aid for disabled and elderly persons in such a way that allows beverage intake in a normal sitting position. In particular, the need to bend down to drink should be avoided by means of the drinking aid. The drinking aid is intended to promote an orthopedically healthy posture while drinking (see

Fig. 9.39 Drinking aid for mobile and stationary use (DE 20208487 U1)

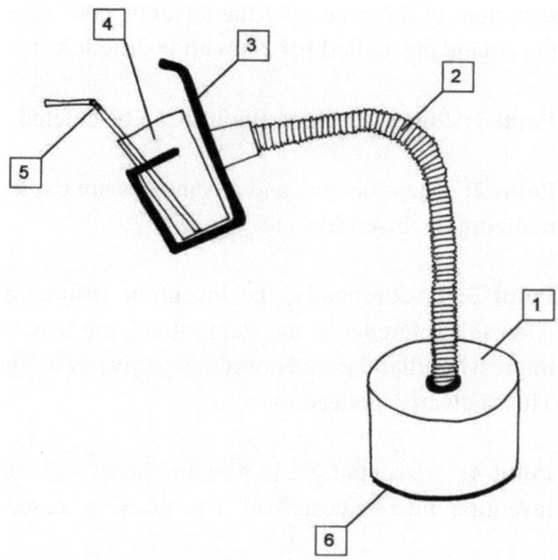

Fig. 9.39, DPMA, https://depatisnet.dpma.de/DepatisNet/depatisnet?action=pdf&docid=DE000020208487U1&xxxfull=1, accessed on 25 April 2026).

The set of claims comprises 12 claims. The first three claims are discussed:

1. Drinking aid in the form of a mobile beverage holder ***for use*** [Point 1] *by disabled, elderly, and persons with restricted motor mobility.* ***The invention is intended in particular, to enable beverage intake in the normal sitting position of a juvenile or adult person. Bending down to a drinking vessel, in the form of a glass, cup, or bottle, is thus eliminated. This ensures an orthopedically correct posture and reduces incorrect loading due to improper sitting positions*** [Point 2]. ***Additionally, the invention ensures a stable stand for the drinking vessel*** [Point 3], *characterized in that the drinking aid* ***essentially*** [Point 4] *consists of five components: the base, the flexible metal articulated tube—hereinafter referred to as gooseneck—the vessel holder, the drinking vessel, and the* ***drinking straw*** [Point 5] ***consists of*** [Point 6]. ***According to the inventive arrangement of the drinking aid, the base stands on a surface, which is usually a table or similar. On the upper side of the base, one end of the flexible metal articulated tube – gooseneck – is attached. At the other end of the flexible – gooseneck – the vessel holder is attached. The vessel holder can thus be swiveled into any desired position*** [Point 7]. *(DPMA, https://depatisnet.dpma.de/DepatisNet/depatisnet?action=pdf&docid=DE000020208487U1&xxxfull=1, accessed on 25 April 2026)*

In principle, a claim should consist of only one sentence. A claim must include the features by which the invention is realized. After describing an individual feature, the advantageous effect or the purpose of the feature and its interaction with other features of the claim may additionally be stated (for example: rollers for …). It is by no means sufficient to describe only the advantageous effects in a claim, since a claim is intended to provide the skilled person with instructions for carrying out the invention.[60]

Point 1: "for use": Superfluous. Can be deleted.

Point 2: The objective and advantages are explained, without specifying the features for realizing the invention.

Point 3: "Additionally, the invention ensures a stable stand for the drinking vessel": After this reference to the stable stand, the features of the invention follow. This gives the impression that the invention's objective is to ensure a stable stand for the drinking aid. This is clearly misleading.

Point 4: "Essentially": In a main claim, all features necessary for the realization of the invention must be described. The wording "essentially" is therefore meaningless.

Point 5: "Drinking straw": A drinking straw is part of the invention? Somewhat odd. It might be more appropriate to use a means-plus-function feature, which could be defined as "means for drawing liquid from the drinking bottle". Concrete embodiments should then be included in the description, for example, a drinking straw and a tube. Alternatively, special embodiments of individual features can be described in the dependent claims.

Point 6: "Consists of": "Consists of" means that the listed features are exhaustive. A workaround could therefore already be achieved by adding an additional feature. By using the word "consists of", imitators can very easily develop a workaround.

Point 7: This passage belongs in the description, but not in the claim.

An alternative wording of the main claim would be:

1. Drinking aid for disabled, elderly, and persons with restricted motor mobility, comprising:

a vessel holder for receiving a drinking glass,
a metal articulated tube, which can be permanently bent in any desired way, and
a means for drawing liquid from the drinking glass.

2. Drinking aid according to claim 1, characterized in that it has a ***base (1) or base plate*** [Point 1] , *which* ***does not have to be permanently connected to a table or the like*** [Point

[60] Section 34 paragraph 4 Patent Act or Article 83 EPC.

2]. *(DPMA, https://depatisnet.dpma.de/DepatisNet/depatisnet?action=pdf&docid=DE0000 20208487U1&xxxfull=1, accessed on 25 April 2026)*

Point 1: "Base or base plate": What is the difference supposed to be between a base and a base plate?

Point 2: "Does not have to be permanently connected to a table or the like": This means that the base plate may or may not have a permanent connection. So both are possible, i.e., any variant is possible. For this reason, this passage is meaningless and can be deleted.

An alternative wording of claim 2 would be:

2. Drinking aid according to claim 1, wherein the drinking aid comprises a base.

3. Drinking aid according to claims 1 and 2, characterized in that the base (1) may be designed as a cylinder, cube, cuboid, cone, or pyramid ***may be formed*** [Point 1]. *(DPMA, https://depatisnet.dpma.de/DepatisNet/depatisnet?action=pdf&docid=DE00 0020208487U1&xxxfull=1, accessed on 25 April 2026)*

Point 1: "May be formed": A claim should describe what is, not what may or could be.

The features of claim 3 can be combined with the feature of claim 2, thereby reducing the number of claims:

Alternative wording for claims 2 and 3:

2. Drinking aid according to claim 1, wherein the drinking aid has a base and/or wherein the base is designed as a cylinder, cube, cuboid, cone, or pyramid.

"And/or" wording: The "or" connection is advantageous because it means that the base does not necessarily have to be a cylinder, cube, cuboid, cone, or pyramid. However, the "or" connection also results in the alternative that the first subordinate clause does not apply, but the second does. This alternative is nonsensical, but in an official or court proceeding, for example in a cancellation procedure, a patent granting procedure, or an infringement procedure, it can easily be corrected by omitting this variant and splitting the claim into two claims. As a general rule, it can be remembered that something can always be omitted. However, nothing can be added.

9.19.36 Example: Pneumatic Drinking Aid

Utility model DE 20107400 U1 (Title: Pneumatic Drinking Aid) describes a device in which drinking a liquid is facilitated by generating an overpressure in a container (see Fig. 9.40 and Table 9.2, DPMA, https://depatisnet.dpma.de/DepatisNet/depatisnet?action =pdf&docid=DE000020107400U1&xxxfull=1, accessed on 25 April 2026).

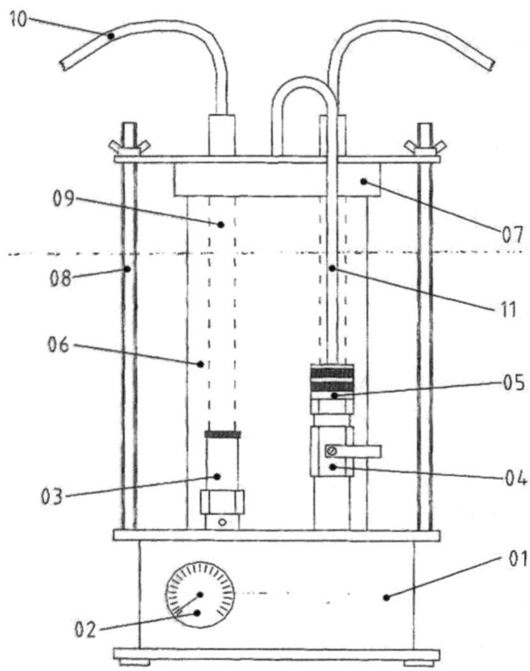

Fig. 9.40 Pneumatic drinking aid (DE 20107400 U1)

Table 9.2 Pneumatic drinking aid—list of reference numerals (DE 20107400 U1)

Components:

01	Compressed air tank
02	Pressure gauge
03	Pressure relief valve
04	Valve
05	Compressed air coupling
06	Vessel
07	Cover
08	Tension rod
09	Tube
10	Drinking tube
11	Compressed air hose

The set of claims reads as follows:

I claim **protection claims** [Item 1] *for the following features of the pneumatic drinking aid* **(position numbers refer to the drawing in the letter dated 19.08.01)** [Item 2]*: (DPMA, https://depatisnet.dpma.de/DepatisNet/depatisnet?action=pdf&docid=DE000020107400U1&xxxfull=1, accessed on 25 April 2026)*

Item 1: In a patent or utility model, the applicant must formulate the protection claims themselves. No infringement court or patent office will select from a sequence of features those that could sensibly form a main claim. This utility model therefore does not provide the applicant with protection against imitators, as there is no fully formulated protection claim.

Item 2: All disclosure components must be included in an application. A reference to correspondence with the patent office is not permitted. As a result, any drawing that is submitted but not filed together with the application documents does not become part of the disclosure of the application.

– Airtight sealing of a container (06) by means of a ***lid*** [Item 1] *(07)* ***made of metal, plastic, ceramic or wood*** [Item 2]. *(DPMA, https://depatisnet.dpma.de/DepatisNet/depatisnet?action=pdf&docid=DE000020107400U1&xxxfull=1, accessed on 25 April 2026)*

Item 1: "Lid": Alternatively, a container could be closed by joining two symmetrical parts together. This feature therefore opens up a workaround solution for an imitator.

Item 2: "made of metal, plastic, ceramic or wood": It makes no sense to list all possible variants; these features can be omitted entirely. The material of the lid also does not appear to be an essential feature of the invention. For this reason, these features can also be omitted. The use of "made of" is disadvantageous, as it implies that the lid consists exclusively of metal or plastic or ceramic or wood. A lid that contains both wood and metal components would result in an object that does not fall within the scope of protection of the utility model. A wording such as "lid comprising metal and/or plastic and/or …" avoids this restriction of the scope of protection.

– Application of compressed air to the container (06). (DPMA, https://depatisnet.dpma.de/DepatisNet/depatisnet?action=pdf&docid=DE000020107400U1&xxxfull=1, accessed on 25 April 2026)

A process is described. However, the present claim is a device claim. It would therefore be sensible to use a feature such as: "wherein an increased pressure exists within the container". However, the "increased pressure" should be further specified, for example by: "…, in particular a pressure greater than …".

The tubes (10) are taken into the mouth by a number of persons to be determined (depending on the expansion stage of the pneumatic drinking aid). When the liquid is subjected to compressed air, it enters the mouths of the persons within a very short time. (DPMA, https://depatisnet.dpma.de/DepatisNet/depatisnet?action=pdf&docid=DE000020107400U1&xxxfull=1, accessed on 25 April 2026)

This passage can be included in the description. It is not suitable for a claim.

– Depending on the expansion stage of the pneumatic drinking aid, the pressure vessel (01) is filled using a commercially available or an FPE-modified hand pump, CO2 cartridges (or other propellants), compressed air cylinders, compressors, or chemical

reactions. (DPMA, https://depatisnet.dpma.de/DepatisNet/depatisnet?action=pdf&docid=DE000020107400U1&xxxfull=1, accessed on 25 April 2026)

Providing a pressure vessel to generate the pressure within the container and filling the pressure vessel are probably not inventive. These features can be included in the description.

– To prevent injury to persons, a pressure relief valve (03) is attached to the pressure vessel (01), which reduces the pressure to a tolerable level. Depending on the expansion stage of the pneumatic drinking aid, the pressure relief valve (03) must be adjusted. For a 1.5 L pressure vessel (01), about 0.8 to 1.4 bar is common, depending on the experience of the persons concerned.—To allow the pressure to be monitored according to the experience of the persons concerned, a pressure gauge (02) is attached to the pressure vessel (01). This can be omitted depending on the expansion stage of the pneumatic drinking aid.—The release of air from the pressure vessel (01) into the container (06) is effected by a valve (04). Depending on the expansion stage of the pneumatic drinking aid, this can be controlled pneumatically, electronically, or mechanically.

– To remove the lid (07) from the container (06) (for cleaning the pneumatic drinking aid), the pressure hose (11) can be detached from the pressure vessel (01). For this purpose, a compressed air coupling (05) is provided. Depending on the expansion stage of the pneumatic drinking aid, this can be replaced by a plug, clamp, or screw connection. (DPMA, https://depatisnet.dpma.de/DepatisNet/depatisnet?action=pdf&docid=DE000020107400U1&xxxfull=1, accessed on 25 April 2026)

At least in part, the state of the art is explained. These passages should be included in the description.

An alternative wording of a main claim could be as follows:

1. Drinking aid for supplying a person with a liquid, comprising:

a container that can be sealed in a pressure-tight manner, wherein a liquid is contained in the container, which is forced through a tube (10) by pressure in the container, thereby supplying the person with the liquid through the tube (10).

9.19.37 Example: Tableware for Children and Seniors

The utility model publication DE 20111403 U1 (Title: Tableware for babies/children, disabled persons, and seniors) describes a learning aid for eating or drinking, which, when secured, prevents it from being thrown down or knocked over.

The set of claims comprises four claims, none of which is formulated as a complete sentence. Such a formulation of the claims is incorrect. Each claim must be expressed in a complete sentence.

The "protective claims" are as follows:

Tableware, ***consisting of*** [Point 1] ***flat and deep*** [Point 2] *plates, bowls, cup, mug, and glass holders, for babies/children, disabled persons, and seniors, which is* **secured with suction cups** [Point 3] *to high chairs, tables, or other smooth surfaces,*

characterized in that, (DPMA, https://depatisnet.dpma.de/DepatisNet/depatisnet?action=pdf&docid=DE000020111403U1&xxxfull=1, accessed on 25 April 2026)

Point 1: "consisting of": is an exhaustive enumeration. A tableware set that additionally included a saucer would therefore not infringe the utility model.

Point 2: "flat and deep": that is, any type of plate, so this feature can be omitted, as it is obviously not essential.

Point 3: "Suction cups": The suction cups are described before the "characterized in that" clause. This implies that the suction cups are not intended to constitute the inventive step. This is likely an error.

An alternative formulation of the main claim would be:

1. Tableware, in particular a plate, a bowl, a cup, a mug, or a glass holder, for a child, a disabled person, or an elderly person, wherein a suction cup is arranged on the tableware for securing the tableware to a high chair, a table, or a smooth surface.

1. that this tableware and the beverage holders, on the base or the standing surfaces, have one or more elevations with round or square holes or recesses, into which a suction cup or several suction cups are inserted/secured accordingly. (DPMA, https://depatisnet.dpma.de/DepatisNet/depatisnet?action=pdf&docid=DE000020111403U1&xxxfull=1, accessed on 25 April 2026)

An alternative claim formulation would be:

2. Tableware according to claim 1, wherein the tableware has an elevation on its underside, and the suction cup is arranged on the elevation.

2. that this tableware and the beverage holders, in a double or thick base or the standing surfaces, have one or more round or square holes or recesses, into which a suction cup or several suction cups are inserted/secured accordingly. (DPMA, https://depatisnet.dpma.de/DepatisNet/depatisnet?action=pdf&docid=DE000020111403U1&xxxfull=1, accessed on 25 April 2026)

An alternative claim formulation would be:

3. Tableware according to one of claims 1 or 2, wherein the tableware has a recess or cavity on its underside for receiving the suction cup.

3. that on the base or the standing surface of the tableware and the beverage holders, one or more suction cups are firmly attached. (DPMA, https://depatisnet.dpma.de/DepatisNet/depatisnet?action=pdf&docid=DE000020111403U1&xxxfull=1, accessed on 25 April 2026)

An alternative claim formulation would be:

4. Tableware according to any of the preceding claims, wherein the suction cup is arranged on the tableware in a detachable or non-detachable manner.

4. that the tableware should be made from all, as far as technically possible, common materials. (DPMA, https://depatisnet.dpma.de/DepatisNet/depatisnet?action=pdf&docid =DE000020111403U1&xxxfull=1, accessed on 25 April 2026)

This claim can be deleted, as it does not provide a precise description of a subject to be protected if all possible variations are claimed without specifying them.

9.19.38 Example: Packaging with a Secondary Use as a Children's Toy[61]

The utility model publication DE 20 2017 006 000 U1 (Title: Packaging with a secondary use as a children's toy) addresses the technical problem of designing packaging so that, in addition to its function as packaging, it can also be used as a children's toy. Here, after unpacking the contents, the packaging is intended to take on a special play form (see Figs. 9.41 and 9.42 and Table 9.3, DPMA, https://depatisnet.dpma.de/DepatisNet/depatisnet?action=pdf&docid=DE202017006000U1&xxxfull=1, accessed on 25 April 2026).

The sole claim reads:

1. __A__ [Point 1] packaging with a secondary use as a children's toy, characterized in that after removal of the original contents of the packaging, it, by virtue of its special shape, is suitable as a children's toy for one or more children's play themes. (DPMA, https://depatisnet.dpma.de/DepatisNet/depatisnet?action=pdf&docid= DE202017006000U1&xxxfull=1, accessed on 25 April 2026)

Point 1: "a": unnecessary, can be omitted.

It is questionable whether the claim is novel. After all, any box can in some way be used as a toy. The claim is also formulated as a problem to be solved. Features that enable the corresponding use have not been included, but it is simply required that the packaging be suitably designed. This raises the suspicion that the invention has not actually been made, since no features could be specified that would lead to the realization of the invention.

The claim could potentially be directed to a specific children's play theme, for example, a medieval castle or a Wild West fort. Alternatively, consideration should be given

[61] Section 1, paragraph 3, no. 3 of the Patent Act deals with games: "The following in particular shall not be regarded as inventions within the meaning [of the Patent Act] … 3. Schemes, rules, and methods for performing mental acts, playing games, or doing business, and programs for computers." Game rules therefore cannot be patented. However, it is possible to patent a concept by which a game can be integrated into another object, as in this case, into packaging.

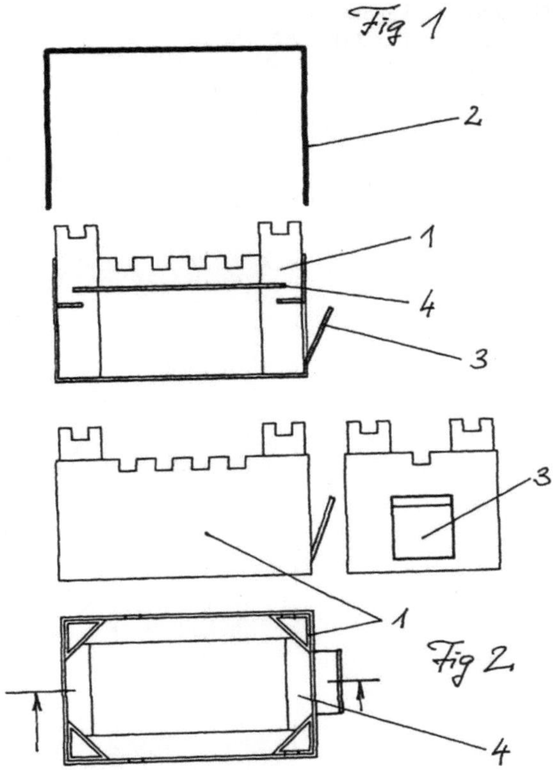

Fig. 9.41 Packaging with a secondary use Figs. 1 and 2 (DE 20 2017 006 000 U1)

to what an abstract feature of the invention could look like. At the very least, a feature should be included that distinguishes it from conventional packages in the prior art, since these can already be used as children's toys.

Proposal for an amended set of claims:

1. Packaging for use as a children's toy, wherein the packaging has a wall with a wall edge that does not run continuously horizontally to the base of the packaging, but in particular is stepped, convex, or concave, or wherein this wall comprises a recess, wherein this wall is part of a children's toy.

2. Packaging according to claim 1, wherein the wall represents a depiction of a medieval castle or a Wild West fort.

9.19.39 Example: Flexible Mount for a Children's Toy

Utility model DE 20 2007 001 551 U1 (Title: Flexible Mount for Children's Toy, which is mounted on strollers) addresses the technical problem of providing a toy holder that can be mounted on a stroller. The inventive toy holder is particularly characterized by

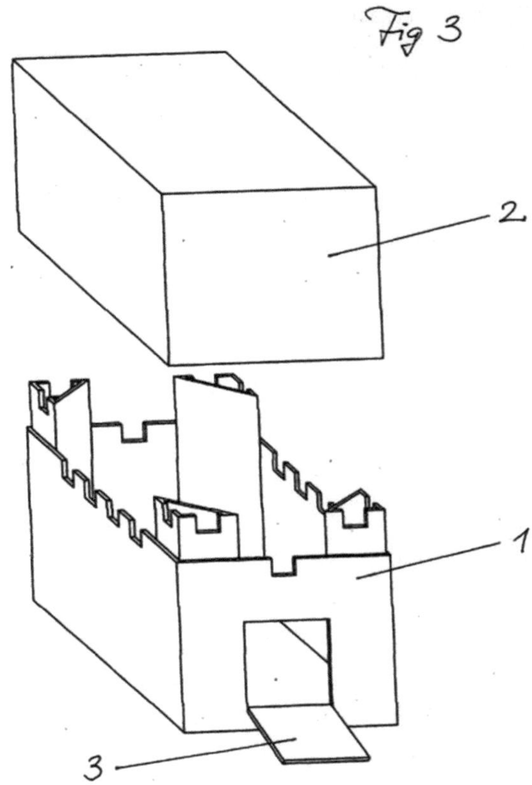

Fig. 9.42 Packaging with a secondary use Fig. 3 (DE 20 2017 006 000 U1)

Table 9.3 Packaging with a secondary use—list of reference numerals (DE 20 2017 006 000 U1)

Reference list

1	Lower part
2	Upper part
3	Fold-out elements (here: drawbridge)
4	Additional elements (here: battlement)

the fact that the children's freedom of movement is not restricted by the toy holder (see Figs. 9.43 and 9.44, DPMA, https://depatisnet.dpma.de/DepatisNet/depatisnet?action=pdf&docid=DE202007001551U1&xxxfull=1, accessed on 25 April 2026).

The set of claims reads as follows:

1. Flexible articulated tube mount ***(gooseneck, Fig. 3a)*** [Point 1] *for attaching stroller toys, in particular a toy steering wheel (Fig. 3b),* ***to the stroller frame*** [Point 2] *(Fig. 2d)* ***in the direction of travel*** [Point 3]. *(DPMA, https://depatisnet.dpma.de/DepatisNet/depatisnet?action=pdf&docid=DE202007001551U1&xxxfull=1, accessed on 25 April 2026)*

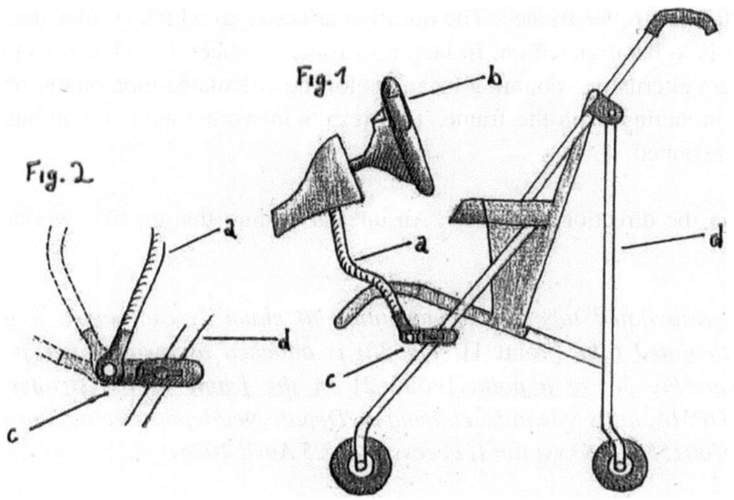

Fig. 9.43 Flexible mount for a children's toy Figs. 1 and 2 (DE 20 2007 001 551 U1)

Fig. 9.44 Flexible mount for a children's toy Fig. 3 (DE 20 2007 001 551 U1)

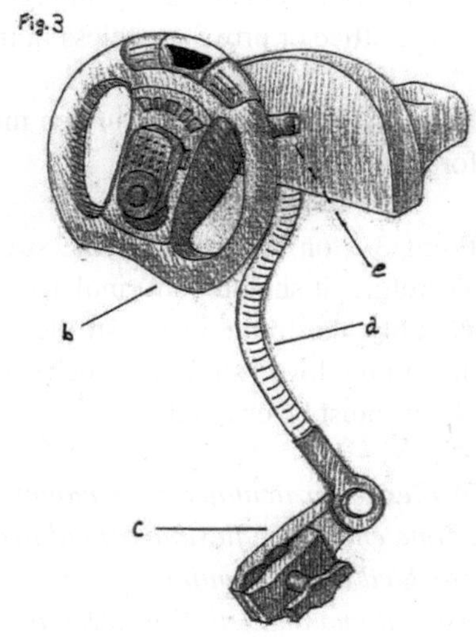

Point 1: "(gooseneck, Fig. 3a)": In a claim, nothing except reference signs should be placed in parentheses, as parentheses signal that the enclosed text is of lesser importance. Insignificant features should not be included in a claim at all. Furthermore, references to drawings should be omitted in a claim.

Point 2: "to the stroller frame": The question arises as to which stroller the articulated tube mount is to be arranged on. In fact, no stroller has been introduced so far. In addition, there are alternative mounting locations for the articulated tube mount on a stroller, apart from mounting it on the frame. This results in workarounds for imitators, which should be prevented.

Point 3: "in the direction of travel": An unclear feature that enables workarounds for imitators.

2. Flexible articulated tube mount according to claim 1, characterized in that ***the flexible articulated tube*** [Point 1] *(Fig. 3a) is attached by means of a quick-release screw clamp (Fig. 3c)* ***at a point*** [Point 2] ***on the frame of the stroller*** [Point 3] *(Fig. 1d). (DPMA, https://depatisnet.dpma.de/DepatisNet/depatisnet?action=pdf&docid=DE202007001551U1&xxxfull=1, accessed on 25 April 2026)*

Point 1: "the flexible articulated tube": An articulated tube has not yet been introduced. Therefore, the indefinite article should be used. Possibly, the articulated tube refers to the articulated tube mount or a part of the articulated tube mount. The applicant must formulate precisely and avoid any ambiguity. Unclear claims regularly do not stand up before a patent office or prove worthless in infringement proceedings.

Point 2: "at a point": A surface mounting would already be a permissible workaround for an imitator.

Point 3: "on the frame of the stroller": A stroller frame has not yet been introduced. Therefore, it should be formulated with an indefinite article: "on a frame of the stroller". Possibly, the frame of the stroller refers to the stroller frame of claim 1, in which case the terminology "stroller frame" should be maintained. Consistency of terms in the set of claims must be ensured.

3. Flexible articulated tube mount according to ***claim 1*** [Point 1] *characterized in that at one end of the flexible articulated tube (Fig. 3a) a* ***screw adapter*** [Point 2] *(Fig. 3e) is provided as a mounting option for a toy steering wheel (Fig. 3b). (DPMA, https://depatisnet.dpma.de/DepatisNet/depatisnet?action=pdf&docid=DE202007001551U1&xxxfull=1, accessed on 25 April 22026)*

Point 1: The back-reference is limited to claim 1. A combination of claims 1, 2, and 3 is therefore not disclosed. Such a combination of features could not be used to distinguish from the prior art. The following back-reference is recommended to enable all claim combinations: "… according to one of claims 1 or 2, …".

Point 2: "Screw adapter": What is a screw adapter? Here, a feature description of the following type could be provided: "... screw adapter, wherein the screw adapter comprises: ... for ...".

4. Flexible articulated tube mount according to ***claim 1*** [Point 1], *characterized in that the flexible articulated tube mount (gooseneck)* ***at the end*** [Point 2] *also has a carabiner hook for attaching hanging toys. (DPMA, https://depatisnet.dpma.de/DepatisNet/depatisnet?action=pdf&docid=DE202007001551U1&xxxfull=1, accessed on 25 April 2026)*

Point 1: Claim 4 refers only to the main claim. A combination of claims 1, 2, and 4 or claims 1, 3, and 4 or claims 1, 2, 3, and 4 is therefore not possible. The following back-reference is recommended to enable all claim combinations: "... according to one of the preceding claims ...".

Point 2: "at the end": A workaround for an imitator would be to arrange the carabiner hook not at the end, but somewhat away from the end of the articulated tube mount.

Proposal for an improved set of claims:

1. Flexible articulated tube mount for arranging a children's toy device, in particular a toy steering wheel, on a stroller.
2. Articulated tube mount according to claim 1, wherein the articulated tube mount is detachably fastened to the stroller by means of a quick-release screw clamp.
3. Articulated tube mount according to one of claims 1 or 2, wherein a screw adapter for attaching a toy steering wheel is arranged at one end of the articulated tube mount, wherein the screw adapter is connected to the toy steering wheel by screwing, and wherein the screw adapter is connected to the articulated tube mount.
4. Articulated tube mount according to one of the preceding claims, wherein the articulated tube mount has a carabiner hook for attaching a toy.

9.19.40 Example: Scooter Toy for Children

The utility model publication DE 20 2006 006 537 U1 (Title: Scooter Toy for Children, Sporting Equipment) addresses the technical problem of providing an ice sled scooter as a winter sports device for movement on ice. For this purpose, a scooter with runners for gliding is proposed. The scooter enables ice skating by pushing off from the ground. Advantageously, shoes with spikes can assist with pushing off (see Fig. 9.45, DPMA, https://depatisnet.dpma.de/DepatisNet/depatisnet?action=pdf&docid=DE202006006537U1&xxxfull=1, accessed on 25 April 2026).

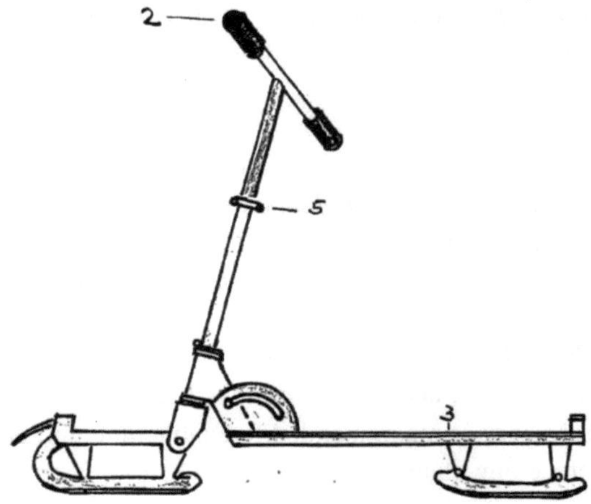

Fig. 9.45 Scooter toy for children (DE 20 2006 006 537 U1)

Claims 1 to 4 are as follows:

1. Ice sled scooter, characterized in that it has a ***standing surface*** [Point 1] *and a scooter handlebar. (DPMA, https://depatisnet.dpma.de/DepatisNet/depatisnet?action=pdf&docid=DE202006006537U1&xxxfull=1, accessed on 25 April 2026)*

In claim 1, one would expect a description that the scooter is suitable for gliding over ice and which features it possesses for this purpose. Instead, the features of a kick scooter from the prior art are described, namely standing surface and handlebar. This claim can be interpreted to cover any kick scooter and is therefore not legally valid.

Point 1: "Standing surface": What does "standing surface" mean? Is it a surface so that the scooter can be parked standing up, or so that a person can stand on a surface of the scooter?

2. Ice sled scooter according to claim 1, characterized in that **runners can be mounted** [Point 2] *under the standing surface and under the handlebar. (DPMA,* https://depatisnet.dpma.de/DepatisNet/depatisnet?action=pdf&docid=DE202006006537U1&xxxfull=1, *accessed on 25 April 2026)*

Point 2: "Runners can be mounted": In principle, runners can be mounted on any kick scooter. The subject matter of claim 2 is therefore not novel. It would be different if it were claimed that "runners are mounted". Such a scooter is not a conventional kick scooter and could therefore be novel and inventive.

3. Ice sled scooter ***according to claim 1. and 2.*** *[Point 1], characterized in that the handle grip, handlebar, handlebar joint, and standing surface are made of steel, aluminum, plastic, carbon, wood, or a combination of these materials. (DPMA,* https://depatisnet.dpma.de/DepatisNet/depatisnet?action=pdf&docid=DE202006006537U1&xxxfull=1, *accessed on 25 April 2026)*

Point 1: "according to claim 1 and 2": This means that only the combination of claims 1, 2, and 3 is disclosed. A combination of claims 1 and 3 (without the features of claim 2) is excluded. An advantageous formulation of the dependencies, which allows all combinations, would be: "according to one of claims 1 or 2".

The features of claim 3 describe common materials for manufacturing a kick scooter. The summary of the invention should explain how these materials or a combination of these materials lead to particular effects, for example, by providing a better grip, making the kick scooter less prone to damage, or lighter and thus easier to transport, etc.

4. Ice sled scooter ***according to claim 1. and 2. and 3.*** [Point 1], *characterized in that* ***the handlebar joint contains a slot, which enables the handlebar, which can be extended up to 90 degrees from the standing surface during use, to be folded*** [Point 2]. *(DPMA, https://depatisnet.dpma.de/DepatisNet/depatisnet?action=pdf&docid=DE202006006537U1&xxxfull=1, accessed on 25 April 2026)*

Point 1: "according to claim 1 and 2 and 3": This means that only the combination of the features of claims 1, 2, 3, and 4 is possible. All other combinations, for example, combining the features of claims 1 and 4 into a new main claim, are excluded. A better formulation of the dependency would be: "... according to one of the preceding claims ..."

Point 2: The claim is unclear and therefore not executable by a person skilled in the art.[62]

9.19.41 Example: Mop

The utility model publication DE 20 2018 005 521 U1 (Title: Mop) addresses the technical problem of providing a mop with a large dirt-absorbing surface (see Fig. 9.46, DPMA, https://depatisnet.dpma.de/DepatisNet/depatisnet?action=pdf&docid=DE202018005521U1&xxxfull=1, accessed on 25 April 2026).

The set of claims comprises two claims:

1. Mop characterized in that it consists of a helical adapter ***(Fig. 2)*** [Point 1] ***consists*** [Point 2]. [Point 3] *The adapter* ***can*** *be made of plastic, metal, or wood* ***manufactured*** [Point 4]. *(DPMA, https://depatisnet.dpma.de/DepatisNet/depatisnet?action=pdf&docid=DE202018005521U1&xxxfull=1, accessed on 25 April 2026)*

Point 1: Claims should not contain references to the description or drawings.

[62] Section 34 paragraph 4 Patent Act or Article 83 EPC.

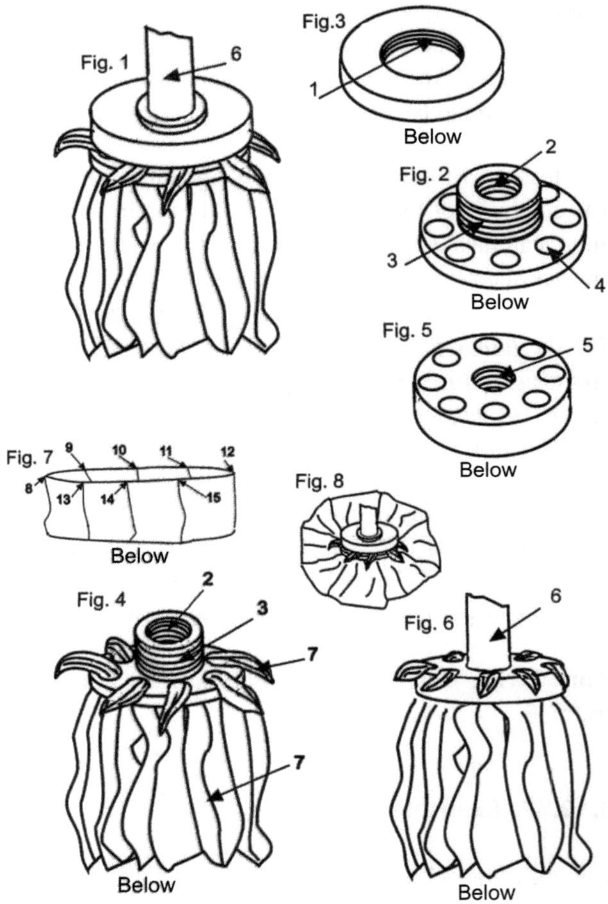

Fig. 9.46 Mop (DE 20 2018 005 521 U1)

Point 2: The verb "consists" is limiting. It should only be used if the invention is also seen in the fact that no additional elements are included. Otherwise, the verbs "comprise" or "include" are always preferable.

Point 3: A claim corresponds to a single sentence.

Point 4: "can … be manufactured": A claim should describe what is, not what can or could be. Therefore, it would be better to formulate: "The adapter comprises plastic, metal and/or wood."

2. Mop according to claim 1, characterized in that the adapter (Fig. 5 and Fig. 6) ***does not have a helical, but a single-stage form*** [Point 5]. *(DPMA,* https://depatisnet.dpma.de/DepatisNet/depatisnet?action=pdf&docid=DE202018005521U1&xxxfull=1, *accessed on 25 April 2026)*

Point 5: Unclear passage. As a minimum requirement, a claim must be understandable.

9.19.42 Example: Foldable Mobile Phone Case on Textiles

The utility model publication DE 20 2017 001 820 U1 (Title: Foldable mobile phone case (smartphone case) on textiles) proposes an inventive garment that prevents a mobile phone from falling out when being removed from the garment (see Figs. 9.47 and 9.48 and Table 9.4, DPMA, https://depatisnet.dpma.de/DepatisNet/depatisnet?action=pdf&docid=DE202017001820U1&xxxfull=1, accessed on 25 April 2026).

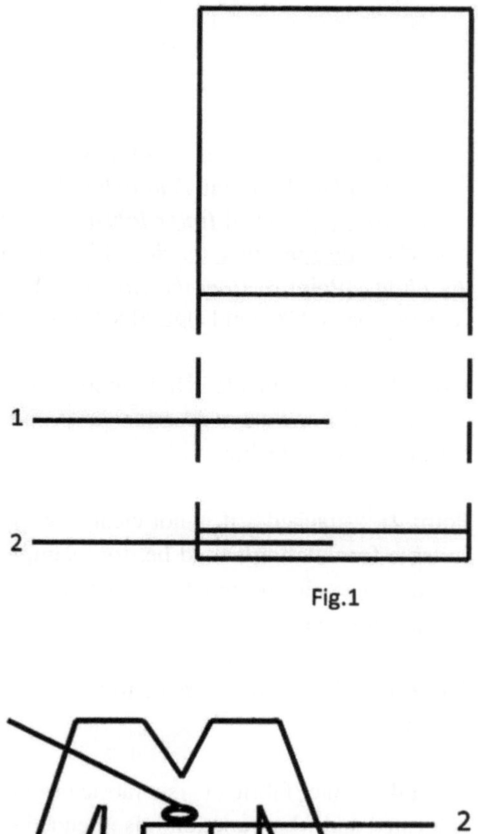

Fig. 9.47 Foldable mobile phone case on textiles Fig. 1 (DE 20 2017 001 820 U1)

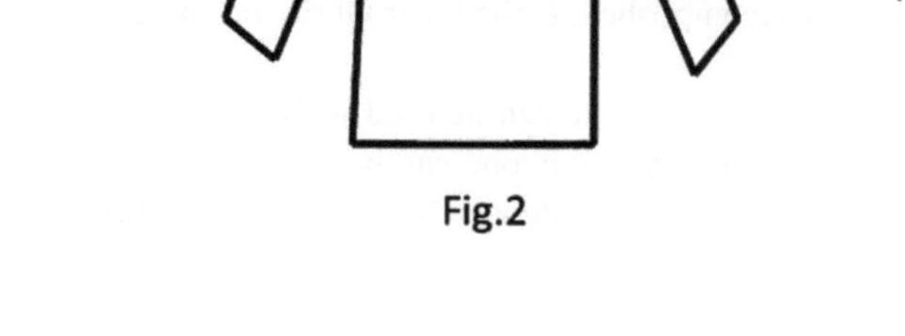

Fig. 9.48 Foldable mobile phone case on textiles Fig. 2 (DE 20 2017 001 820 U1)

Table 9.4 Foldable mobile phone case on textiles—list of reference signs (DE 20 2017 001 820 U1)

Reference list

Fig. 1

1	Inner fabric
2	Closure device

Fig. 2

1	Closure device for opening and closing
2	Closure device for opening and closing to store a cell phone
3	Outer fabric
4	Attachment of the foldable cell phone case

The set of claims reads as follows:

1. ***The foldable*** [Point 1] ***attached*** [Point 2] *mobile phone case (smartphone case) on textiles* ***Fig. 1*** [Point 3] ***Inner fabric (1) is a fabric or textile film*** [Point 4] ***the user can read through the inner fabric (1)*** [Point 5]. ***The phone can be used without removing the phone*** [Point 6] *from the case. (DPMA,* https://depatisnet.dpma.de/DepatisNet/depatisnet?action=pdf&docid=DE202017001820U1&xxxfull=1*, accessed on 25 April 2026)*

Point 1: "The foldable": It is more elegant to first name the object, namely "mobile phone case", and then describe its properties, namely that the mobile phone case is designed to be foldable.

Point 2: "attached": It is not clear how or where the mobile phone case is attached. A suitable formulation would be, for example: "mobile phone case for receiving a mobile phone", wherein the mobile phone case is attached to a garment, in particular a shirt, T-shirt, or sweater.

Point 3: "Fig. 1": A reference to the drawings or the description should not be included in a claim.

Point 4: "Inner fabric (1) is a fabric or textile film": It is not clear where the inner fabric is located and what function it is intended to serve.

Point 5: "the user can read through the inner fabric (1)": It is not described how this task is accomplished. Is the inner fabric transparent or does it have openings?

Point 6: "The phone can be used without removing the phone": It could be formulated more clearly that the phone can be used without being removed from the mobile phone case. However, it should also be described how this is achieved.

An improved claim could read as follows:

1. Garment, in particular a shirt, trousers, T-shirt, or sweater, comprising:

a mobile phone case for receiving a smartphone, wherein the mobile phone case has a flap, and the flap can be opened to view and operate the display of the smartphone.

9.19.43 Example: Luminous Band for Baby and Children's Drinking Bottles

The utility model DE 20 2021 000 262 U1 (Title: Luminous band for baby and children's drinking bottles) addresses the technical problem of quickly finding a baby bottle in the dark, for example at night, in order to give a crying baby the bottle (see Fig. 9.49, DPMA, https://depatisnet.dpma.de/DepatisNet/depatisnet?action=pdf&docid=DE202021000262U1&xxxfull=1, accessed on 25 April 2026).

The set of claims reads as follows:

1. Luminous band for baby and children's drinking bottles, characterized in that in a dark environment, the drinking vessel equipped with it can be easily found and grasped due to the afterglow effect of the band-shaped ring. (DPMA, https://depatisnet.dpma.de/DepatisNet/depatisnet?action=pdf&docid=DE202021000262U1&xxxfull=1, *accessed on 25 April 2026)*

The claim actually only describes the objective: "that in a dark environment, the drinking vessel equipped with it can be easily found and grasped due to the afterglow effect of the band-shaped ring." Only by specifying the subject matter of the claim, "luminous band," is the solution to the problem implied. (DPMA, https://depatisnet.dpma.de/DepatisNet/depatisnet?action=pdf&docid=DE202021000262U1&xxxfull=1, accessed on 25 April 2026)

The main claim claims a luminous band and describes its use on a drinking bottle. The main claim is obviously not new, since luminous bands already exist and can natu-

Fig. 9.49 Luminous band for baby and children's drinking bottles (DE 20 2021 000 262 U1)

rally be used in this way. The main claim should therefore be directed to a drinking bottle, in particular a baby bottle, in order to potentially be novel and inventive.

The main claim is also too narrowly formulated, as it claims only a very specific embodiment of the invention. The fundamental invention is that a drinking bottle is self-illuminating so that it can be found in the dark. The drinking bottle must therefore have a section that glows in the dark. This could also be a ring of LEDs or any other light source in any form or design.

An improved main claim would be:

1. Drinking bottle, in particular a baby bottle, comprising a luminous section so that the drinking bottle is visible in the dark.

9.20 Tips for Drafting Claims

Drafting claims regularly presents a major hurdle when preparing a patent or utility model application. Therefore, tips are provided to help achieve an appropriate set of claims.

Tip: The main claim should include only those features of an invention that are absolutely necessary for its realization.

The main claim should not contain unnecessary features. Only those features required to solve the technical problem should be included in the main claim or in an independent claim. Otherwise, the scope of protection claimed will be too narrow.

Tip: Claims should be formulated clearly.

An artificially legalistic style should be avoided. Claims must be clear, unambiguous, and precise. Ambiguous wording always works to the detriment of the applicant.

Tip: In the main claim and independent claims, terms should be used that do not exclude similar, obvious implementations. Means-plus-function claim formulations are often useful.

The classic example of an overly narrow formulation is the nail for fastening a device. Fastening with a hook-and-loop fastener might be obvious and could also serve to realize the invention. However, the hook-and-loop fastener would already be a legitimate workaround for an imitator. It would be more appropriate to use the term “fastening means” or “means for fastening”, which would encompass a nail, a dowel, and a hook-and-loop fastener. In the description, a concrete example should always be included for such means-plus-function terms. This preempts the objection of insufficient enablement.

Concrete embodiments, such as a nail and a hook-and-loop fastener, can additionally be mentioned in the dependent claims.

Tip: Consistency of terminology in the set of claims must be ensured.

A “joint pipe holder” is not identical to a “holder”. A “joint pipe” is not automatically a part of a joint pipe holder. Terms must be used precisely in a set of claims. If a “joint pipe holder” has been introduced, this term should be used consistently throughout. If the term “joint pipe” is used, it should be described that, and which part of the joint pipe holder is the joint pipe.

Tip: Claims should not contain references to the description or the drawings.

Claims should not contain references to drawings. A claim should be understandable on its own.

Tip: The dependent claims may describe specific embodiments of the invention.

Dependent claims serve to further develop the technical teaching of the main claim and to describe specific embodiments of the essential features of the invention. However, one should refrain from describing features that have no chance of distinguishing a main claim from the prior art. Features that cannot possibly establish novelty and inventive step do not belong in a dependent claim.

Tip: The dependencies of the dependent claims should always refer to all preceding claims. A typical formulation: “... according to any one of the preceding claims...”. If claim 10 refers back to a parallel claim (for example, claim 4), it can be formulated as: “according to any one of claims 4 to 9...”.

The dependencies of the dependent claims should be formulated so that no meaningful combination is excluded.

Informing the Public: The Abstract

10

The summary must state the title of the application and a brief description of the invention in a single sentence. In addition, a drawing must be attached to the summary, from which the essential elements of the invention can be discerned. For a utility model application, a summary is not required.

The summary serves solely to provide technical information to the public. The summary does not form part of the description of the invention. It is therefore not possible, in an official grant procedure, to rely on a feature that is contained exclusively in the summary. Therefore, features of the invention should never be described solely in the summary and not additionally in the description or the claims.

If the patent application includes drawings, the applicant attaches a drawing to the summary. In practice, it is advisable to select a drawing that contains those reference numerals that are also used in the text of the summary.

10.1 Example: Milling Tool for Milling Fiber-Reinforced Plastics

Application DE 10 2019 109 692 A1 (Title: Milling tool for milling fiber-reinforced plastics) has the following main claim:

1. Milling tool (1) for milling fiber-reinforced plastics, with a shank section (12) and a cutting section (11), wherein the cutting section (11) comprises several webs (31, 32) separated by flutes (2), wherein at least one web (31) has a lateral surface with a plurality of teeth (4), and wherein at least one web (32) has a lateral surface with a plurality of cutting edges (5), and wherein the cutting edges (5) are arranged on the web (32) in a spiral segment shape, each at a different helix angle (αn) relative to one

T. H. Meitinger, *Patent Made Simple: How to File Without a Lawyer*,
https://doi.org/10.1007/978-3-662-73096-6_10

another. (DPMA, https://depatisnet.dpma.de/DepatisNet/depatisnet?action=pdf&docid=DE102019109692A1&xxxfull=1, *accessed on 25 April 2026)*

The main claim was used as the summary:

Summary: [Point 1] *Milling tool 1 for milling fiber-reinforced plastics, with a shank section 12 and a cutting section 11, wherein the cutting section 11 comprises several webs 31, 32 separated by flutes 2, wherein at least one web 31 has a lateral surface with a plurality of teeth 4, and wherein at least one web 32 has a lateral surface with a plurality of cutting edges 5, and wherein the cutting edges 5 are arranged on the web 32 in a spiral segment shape, each at a different helix angle* α_n *relative to one another.*[1]

It is common practice to use the main claim as the text of the summary.

Point 1: It is not necessary to place "Summary:" at the beginning of the summary text.

10.2 Example: Luggage Carrier for a Single-Track Vehicle

In patent application DE 19853172 A1 (Title: Luggage carrier for a single-track vehicle), a very detailed and comprehensive summary is provided:

The invention relates to a luggage carrier for single-track vehicles, in particular for bicycles with a tubular frame or similar structural elements. [Point 1] *The luggage carrier has a luggage support 5 arranged on a support element 4 and a clamping bracket for attachment to the vehicle.* ***The known luggage carriers have the disadvantage that the type of attachment poses a latent risk of overloading both for the vehicle and the luggage carrier; moreover, it does not provide sufficient protection against rotation of the luggage carrier, including the luggage, around the clamping point, so that driving stability and thus road safety can be significantly impaired; in addition, the luggage carrier cannot be positioned adequately.*** [Point 2] ***The object of the invention is to provide a luggage carrier for single-track vehicles that offers protection against excessive luggage loads as well as against unwanted rotation, and that allows its positioning to be adapted to different conditions.*** [Point 3] *According to the invention, an overload protection prevents mechanical damage to the luggage carrier or the vehicle; it is designed so that it automatically yields when a predetermined nominal load is exceeded. In addition, supporting elements prevent lateral rotation, whereby an adjusting element 13 allows alignment with the vehicle centerline and the support bracket 16 allows adjustment to the horizontal …*[Point 4]. (DPMA, https://depatisnet.dpma.de/DepatisNet/depatisnet?action=pdf&docid=DE000019853172A1&xxxfull=1, accessed on 25 April 2026)

[1] DE 10 2019 109 692 A1, DPMA, https://depatisnet.dpma.de/DepatisNet/depatisnet?action=pdf&docid=DE102019109692A1&xxxfull=1, accessed on 25 April 2026.

Point 1: The passage "The invention relates to…" constitutes the first section of an application. This section does not belong in the summary, but rather in the "Field of the Invention" section of the patent or utility model application.

Point 2: This section of the summary explains the disadvantages of the prior art. This discussion of the prior art does not belong in a summary, but rather constitutes the first section of the "Summary of the Invention."

Point 3: The technical problem addressed by the invention is described. This passage is a central part of the "Summary of the Invention" section. It is neither necessary nor customary to repeat this section in the summary.

Point 4: A summary may contain a maximum of 1,500 characters.[2] A summary should therefore not be overly long. The summary serves solely to provide brief and concise information to the public,[3] which means that it is, strictly speaking, not part of the application documents. It is not possible to use descriptions from the summary to distinguish the main claim or independent claims from documents of the prior art. Therefore, it is not advisable to devote too much effort to preparing the summary.

Although the Patent Act requires that the summary describe the technical field of the invention, the technical problem, the solution to the problem (i.e., the actual invention), and the possible uses of the invention,[4] in practice these points are described in detail in the "Field of the Invention," the "Background of the Invention", and the other parts of the actual patent application. The summary should not contain any description of the invention or its development that is not verbatim in the actual patent or utility model application.

Tip: The main claim can be used as the summary.

By using the main claim as the summary, the requirements of the Patent Act are fulfilled, since the main claim discloses the technical teaching in the relevant technical field. The main claim describes the solution to the technical problem. Typically, a device or a method is claimed, with an additional indication of the application for which the device or method is particularly suitable ("Device for …" or "Method for …"). The main claim therefore generally also describes the principal or a suitable possible use.

---

[2] Section 13(1) Patent Ordinance.

[3] Section 36(2) sentence 1 Patent Act.

[4] Section 36(2) sentence 2 no. 2 Patent Act.

11 Application Forms

The application documents must be accompanied by the relevant application form of the respective patent office. This form must contain information that enables the patent office to at least contact the applicant. In addition, the application must be signed.[1]

11.1 Application for Grant of a German Patent

A German patent grants a monopoly right for an invention throughout the entire territory of the Federal Republic of Germany. A German patent application can be submitted electronically, by post, or by fax. If an application is submitted by fax, it must also be subsequently submitted by post. The advantage of submitting the application by fax is that the date of fax transmission is recognized as the filing date.

The German Patent Office must be able to infer from the application for grant of a patent that a patent is indeed being applied for. In addition, the application must contain information that allows the identity of the applicant to be established or at least enables contact with the applicant.[2]

[1] Rule 41 paragraph 2 letter h EPC.

[2] Section 34 paragraph 3 no. 1 Patent Act.

T. H. Meitinger, *Patent Made Simple: How to File Without a Lawyer*,
https://doi.org/10.1007/978-3-662-73096-6_11

The application form for the grant of a patent can be accessed via the link: "https://www.dpma.de/docs/formulare/patent/p2007.pdf" of the German Patent Office.[3] This is the "P2007" form. In field 1, a postal address must be provided. If the information in field 1 matches that of the applicant (field 3, first box checked), then field 4 does not need to be completed. In field 6, the title of the application must be entered. If a priority is to be claimed, field 9 must be completed. It is also important to pay attention to field 12 and to sign the application (see Figs. 11.1, 11.2 and 11.3).

In addition, the inventor must be named.[4] If the applicant is not the inventor, it must be stated how the right to the invention was transferred to the applicant.[5] The form for naming the inventor can be accessed via the link: "https://www.dpma.de/docs/formulare/patent/p2792.pdf" of the German Patent Office.[6] This is the "P2792" form (see Figs. 11.4 and 11.5).

[3] German Patent and Trade Mark Office (DPMA), Zweibrückenstrasse 12, 80.331 Munich.

[4] The deadline for naming the inventor is 15 months from the filing date (Section 37 paragraph 1 sentence 1 Patent Act).

[5] Section 37 paragraph 1 sentence 2 Patent Act.

[6] German Patent and Trade Mark Office (DPMA), Zweibrückenstrasse 12, 80.331 Munich.

Deutsches
Patent- und Markenamt

An das
Deutsche Patent- und Markenamt
80297 München

P 2007 4.19 1

Antrag auf Erteilung eines Patents **1**

(1) Vordruck nicht für PCT-Verfahren verwenden, siehe Seite 4 und 5

Sendungen des Deutschen Patent- und Markenamts sind zu richten an:

Name, Vorname oder Firma

Straße, Hausnummer / ggf. Postfach

Postleitzahl **Ort** **Staat** (falls nicht Deutschland)

Datum TT MM JJJJ

☐ **TELEFAX** vorab am TT MM JJJJ

(2) Zeichen des Anmelders/Vertreters (max. 20 Stellen) Telefon des Anmelders/Vertreters

(3) Der Empfänger in Feld (1) ist der

☐ Anmelder ☐ Zustellungsbevollmächtigte ☐ Vertreter

ggf. Nr. der Allgemeinen Vollmacht

(4) nur auszufüllen, wenn abweichend von Feld (1)

Anmelder (☐ **weitere Anmelder/vertretungsberechtigte Gesellschafter einer GbR sind auf einem gesonderten Blatt angegeben)**

Name, Vorname oder Firma lt. Handelsregister

Straße, Hausnummer (kein Postfach!)

Postleitzahl **Ort** **Staat** (falls nicht Deutschland)

Handelsregisternummer nur bei Firmen anzugeben

☐ Der Anmelder ist eingetragen im Handelsregister Nr.

beim Amtsgericht

Vertreter

Name, Vorname / Bezeichnung

Straße, Hausnummer

Postleitzahl **Ort** **Staat** (falls nicht Deutschland)

Fig. 11.1 German patent application form—1 (DPMA)

P 2007 4.19 2

(5) soweit bekannt

Anmelder-Nr. Vertreter-Nr.

Zustelladressen-Nr.

(6) siehe Seite 4 und 5

Bezeichnung der Erfindung

IPC-Vorschlag des Anmelders (sofern bekannt)

(7) siehe Erläuterung und Kostenhinweise auf Seite 4 und 5

Sonstige Anträge

☐ **Prüfungsantrag** - Prüfung der Anmeldung **mit** Ermittlung des Standes der Technik (§ 44 Patentgesetz)

☐ **Rechercheantrag** - Ermittlung des Standes der Technik **ohne** Prüfung (§ 43 Patentgesetz)

☐ **Aussetzung** des Erteilungsbeschlusses auf ______ Monate (§ 49 Absatz 2 Patentgesetz)
(Max. 15 Monate ab Anmelde- oder Prioritätstag)

(8)

Erklärungen

☐ **Teilung** ☐ **Ausscheidung** aus der Patentanmeldung → Aktenzeichen der Stammanmeldung

☐ an **Lizenzvergabe** interessiert (unverbindlich)

☐ **Nachanmeldung im Ausland** beabsichtigt (unverbindlich)

(9) siehe auch Seite 4 und 5

☐ Inländische **Priorität** (Datum, Aktenzeichen der Voranmeldung)

☐ Ausländische **Priorität**
(Datum, Land, Aktenzeichen der Voranmeldung; vollständige **Abschrift(en) der ausländischen Voranmeldung(en)** beifügen)

Fig. 11.2 German patent application form—2 (DPMA)

P 2007 4.19 3

(10) Erläuterung und Kostenhinweise siehe Seite 4 und 5

Gebührenzahlung in Höhe von ____________ EUR

Zahlung per Banküberweisung

☐ **Überweisung** *(nach Erhalt der Empfangsbestätigung)*

Zahlungsempfänger:
Bundeskasse Halle/DPMA
IBAN: DE84 7000 0000 0070 0010 54
BIC (Swift-Code): MARKDEF1700

Anschrift der Bank:
Bundesbankfiliale München
Leopoldstr. 234, 80807 München

Zahlung mittels SEPA-Basis-Lastschrift

☐ Ein gültiges SEPA-Basis-Lastschriftmandat (Vordruck A 9530)

☐ liegt dem DPMA bereits vor *(Mandat für mehrmalige Zahlungen)*.

☐ ist beigefügt.

☐ Angaben zum Verwendungszweck (Vordruck A 9532) des Mandats mit Mandatsreferenznummer sind beigefügt.

Wird die Anmeldegebühr nicht innerhalb von drei Monaten nach dem Tag des Eingangs der Anmeldung gezahlt, so gilt die Anmeldung als zurückgenommen!

(11) siehe auch Seite 4 und 5

Anlagen

1. ____________ Vertretervollmacht
2. ____________ Erfinderbenennung (P 2792)
3. ____________ Zusammenfassung (ggf. mit Zeichnung Fig. ____________)
4. ____________ Seite(n) Beschreibung (ggf. mit Bezugszeichenliste)
5. ____________ Seite(n) Patentansprüche
   ____________ Anzahl Patentansprüche
6. ____________ Blatt Zeichnungen
7. ____________ Abschrift(en) der Voranmeldung(en)
8. ____________ Zitierte Nichtpatentliteratur
9. ____________ Anzahl Datenträger
   ☐ für Sequenzprotokoll nach § 11 Absatz 2 Patentverordnung
10. ____________ Seite(n) Angaben zum geographischen Herkunftsort des biologischen Materials gemäß § 34a Patentgesetz
11. ____________

Bitte beachten Sie hinsichtlich der Verarbeitung Ihrer personenbezogenen Daten unser Merkblatt A 9106 „Datenschutz bei Schutzrechtsanmeldungen". Dieses finden Sie unter www.dpma.de: Service – Formulare – Sonstige Formulare – Hinweise zum Datenschutz.

____

(12) **Unterschrift/en (Bei mehreren Anmeldern ohne gemeinsamen Vertreter sind die Unterschriften sämtlicher Anmelder erforderlich)**

____

(13) **Funktion des Unterzeichners**

Bitte beachten Sie die Hinweise auf den nächsten Seiten

Fig. 11.3 German patent application form—3 (DPMA)

Deutsches
Patent- und Markenamt

Erfinderbenennung

Die Erfinderbenennung muss auch erfolgen, wenn der Anmelder selbst der Erfinder ist. Ist der Anmelder Miterfinder, so ist er auch mitzubenennen.

Amtliches Aktenzeichen *(wenn bereits bekannt)*

Platz für **Zeichen des Anmelders/Vertreters**

Bezeichnung der Erfindung *(bitte vollständig)*

Erfinder (1)
Vor- und Zuname
Straße, Hausnummer
Postleitzahl Ort

Erfinder (2)
Vor- und Zuname
Straße, Hausnummer
Postleitzahl Ort

Erfinder (3)
Vor- und Zuname
Straße, Hausnummer
Postleitzahl Ort

Fig. 11.4 Inventor designation—1 (DPMA)

P2792 7.18 2

Erfinder (4)

Vor- und Zuname

Straße, Hausnummer

Postleitzahl Ort

Achtung: bei mehr als vier Erfindern bitte gesondertes Blatt benutzen!

Das Recht auf das Patent ist **auf den Anmelder übergegangen durch**:

(z.B. Erfinder ist/sind der/die Anmelder, Inanspruchnahme aufgrund §§ 6 u. 7 ArbnErfG, Kaufvertrag mit Angabe des Datums, Erbschaft usw.)

Bitte beachten Sie hinsichtlich der Verarbeitung Ihrer personenbezogenen Daten unser Merkblatt A 9106 "Datenschutz bei Schutzrechtsanmeldungen". Dieses finden Sie unter www.dpma.de: Service – Formulare – Sonstige Formulare – Hinweise zum Datenschutz.

Es wird versichert, dass nach Wissen des/der Unterzeichner/s weitere Personen an der Erfindung nicht beteiligt sind.

, den

Eigenhändige Unterschrift des Anmelders oder der Anmelder bzw. des Vertreters.
Bei Firmen genaue, eingetragene Firmenbezeichnung angeben.

Antrag auf Nichtnennung als Erfinder

Nur von denjenigen oben genannten Erfindern auszufüllen, die nach außen hin nicht bekanntgegeben werden wollen (§ 63 Abs. 1 S. 3 PatG).
Der Antrag kann jederzeit widerrufen werden. Ein Verzicht des Erfinders auf Nennung ist ohne rechtliche Wirksamkeit (§ 63 Abs. 1 S. 4 u. 5 PatG).

☐ Es wird beantragt, den bzw. die Unterzeichner dieses Antrags in der oben angegebenen Patentanmeldung als Erfinder nicht öffentlich bekanntzugeben. Die Einsicht in die obige Erfinderbenennung wird nur bei Glaubhaftmachung eines berechtigten Interesses gewährt.

, den

Eigenhändige Unterschrift des Erfinders oder der Erfinder

Fig. 11.5 Inventor designation—2 (DPMA)

11.2 Application for Registration of a Utility Model

The application for registration of a utility model can be accessed via the link: "https://www.dpma.de/docs/formulare/gebrauchsmuster/g6003.pdf" of the German Patent Office.[7] This is the "G6003" form. In field 1, a postal address must be provided. If the information matches that of the applicant (field 3, first box checked), then field 4 does not need to be completed. In field 6, the title of the application (designation of the invention) must be entered. If a priority is to be claimed, field 9 must be completed. It is also important to pay attention to field 12 and to sign the application (see Figs. 11.6, 11.7 and 11.8). Naming the inventor is not required for a utility model.

[7] German Patent and Trade Mark Office (DPMA), Zweibrückenstrasse 12, 80.331 Munich.

Deutsches
Patent- und Markenamt

An das
Deutsche Patent- und Markenamt
80297 München

G6003 2.19 1

Antrag auf Eintragung eines Gebrauchsmusters **2**

(1) **Sendungen des Deutschen Patent- und Markenamts sind zu richten an:**

Name, Vorname oder Firma

Straße, Hausnummer / ggf. Postfach

Postleitzahl **Ort**

Datum TT MM JJJJ

☐ **TELEFAX** vorab am TT MM JJJJ

(2) Zeichen des Anmelders/Vertreters (max. 20 Stellen) | Telefon des Anmelders/Vertreters

(3) Der Empfänger in Feld (1) ist der

☐ Anmelder ☐ Zustellungsbevollmächtigte ☐ Vertreter

ggf. Nr. der Allgemeinen Vollmacht

(4) nur auszufüllen, wenn abweichend von Feld (1)

Anmelder (für weitere Anmelder bitte gesondertes Blatt benutzen)

Name, Vorname oder Firma lt. Handelsregister

Straße, Hausnummer (kein Postfach!)

Postleitzahl **Ort** **Land** (falls nicht Deutschland)

Handelsregisternummer nur bei Firmen anzugeben

☐ Der Anmelder ist eingetragen im Handelsregister Nr.

beim Amtsgericht

Vertreter

Name, Vorname / Bezeichnung

Straße, Hausnummer

Postleitzahl **Ort**

Fig. 11.6 Utility model application form—1 (DPMA)

G6003 2.19 2

(5) soweit bekannt

Anmelder-Nr. ________ Vertreter-Nr. ________

Zustelladressen-Nr. ________

(6) siehe Seite 4 IPC-Vorschlag ist unbedingt anzugeben, sofern bekannt

Bezeichnung der Erfindung

IPC-Vorschlag des Anmelders: ☐☐☐☐☐☐☐ / ☐☐☐☐☐

____

(7) siehe Erläuterung und Kostenhinweise auf Seite 4 und 5

Sonstige Anträge

☐ **Aussetzung** der Eintragung und Bekanntmachung für ________ Monate (§ 8 Absatz 1 Satz 2 Gebrauchsmustergesetz) *(Max. 15 Monate ab Anmelde- bzw. Prioritätstag)*

☐ **Rechercheantrag** - Ermittlung der öffentlichen Druckschriften (§ 7 Gebrauchsmustergesetz)

(8)

Erklärungen

	Aktenzeichen	Anmeldetag TT	MM	JJJJ
☐ **Teilung/Ausscheidung** aus der Gebrauchsmusteranmeldung ➔				
☐ **Abzweigung** aus der Patentanmeldung/dem Patent ➔				

☐ Der Anmelder ist an **Lizenzvergabe** interessiert (unverbindlich)

(9)

Priorität

☐ Inländische Priorität (Datum, Aktenzeichen der Voranmeldung) ➔ ________

☐ Ausländische Priorität (Datum, Land, Aktenzeichen der Voranmeldung) ➔ ________

☐ Ausstellungspriorität (Datum der erstmaligen Zurschaustellung, Ausstellung) ➔ ________

Fig. 11.7 Utility model application form—2 (DPMA)

G6003 2.19 3

(10) siehe Kosten-hinweise auf Seite 4 und 5

Gebührenzahlung in Höhe von ________ EUR

Zahlung per Banküberweisung

☐ **Überweisung** *(nach Erhalt der Empfangsbestätigung)*

Zahlungsempfänger:
Bundeskasse Halle/DPMA
IBAN: DE84 7000 0000 0070 0010 54
BIC (Swift-Code): MARKDEF1700

Anschrift der Bank:
Bundesbankfiliale München
Leopoldstr. 234, 80807 München

Zahlung mittels SEPA-Basis-Lastschrift

☐ Ein gültiges SEPA-Basis-Lastschriftmandat (Vordruck A 9530)

☐ liegt dem DPMA bereits vor *(Mandat für mehrmalige Zahlungen)*.

☐ ist beigefügt.

☐ Angaben zum Verwendungszweck (Vordruck A 9532) des Mandats mit Mandatsreferenznummer sind beigefügt.

Wird die Anmeldegebühr nicht innerhalb von 3 Monaten nach dem Tag des Eingangs der Anmeldung gezahlt, so gilt die Anmeldung als zurückgenommen!

(11) **Anlagen**

1. ________ Seite(n) Beschreibung
2. ________ Seite(n) Schutzansprüche ________ Anzahl Schutzansprüche
3. ________ Blatt Zeichnungen
4. ________ Abschrift(en) der Voranmeldung(en) bei Priorität
5. ________ Abschrift der Voranmeldung bei Abzweigung
6. ________ Vertretervollmacht
7. ________

Bitte beachten Sie hinsichtlich der Verarbeitung Ihrer personenbezogenen Daten unser Merkblatt A 9106 „Datenschutz bei Schutzrechtsanmeldungen". Dieses finden Sie unter www.dpma.de: Service – Formulare – Sonstige Formulare – Hinweise zum Datenschutz.

____
(12) **Unterschrift(en)**

____
(13) **Funktion des Unterzeichners**

Bitte beachten Sie die Hinweise auf den nächsten Seiten

Fig. 11.8 Utility model application form—3 (DPMA)

11.3 Application for Grant of a European Patent

The European Patent Office must be able to discern from the application for grant that a patent is indeed being applied for. In addition, the application must contain information that allows the identity of the applicant to be established, or at least enables contact to be made with the applicant.

The application for grant of a European patent can be obtained from the link "http://documents.epo.org/projects/babylon/eponet.nsf/0/5C683C367A8DFBC7C12577F400449FD8/$FILE/epo_form_1001_10_19_editable.pdf" of the European Patent Office.[8]

In the application, under items 7 and 8, the applicant and their address must be specified. Under item 24, the title of the application (designation of the invention) must be entered. Under item 46, the application must be signed (see Figs. 11.9, 11.10 and 11.11).

[8] European Patent Office (EPO), Bob-van-Benthem-Platz 1, 80.469 Munich.

Europäisches Patentamt
European Patent Office
Office européen des brevets

Antrag auf Erteilung eines europäischen Patents
Request for grant of a European patent
Requête en délivrance d'un brevet européen

☐ Nachreichung von Form 1001 zu einer früher eingereichten Anmeldung nach Regel 40 (1) vom
Form 1001 filed further to a previous application under Rule 40(1) on
Dépôt du formulaire 1001 pour une demande déposée antérieurement au titre de la règle 40(1) en date du

☐ Bestätigung einer bereits durch Fax eingereichten Anmeldung vom / Confirmation of an application already filed by fax on / Confirmation d'une demande déjà déposée par télécopie le — bei / with / auprès de

Nur für amtlichen Gebrauch / For official use only / Cadre réservé à l'administration	
1 Anmeldenummer / Application No. / N° de la demande — MKEY	
2 Tag des Eingangs (Regel 35 (2)) / Date of receipt (Rule 35(2)) / Date de réception (règle 35(2)) — DREC	
3 Tag des Eingangs beim EPA (Regel 35 (4)) / Date of receipt at EPO (Rule 35(4)) / Date de réception à l'OEB (règle 35(4)) — RENA	
4 Anmeldetag / Date of filing / Date de dépôt	

5 Es wird die Erteilung eines europäischen Patents und gemäß Artikel 94 die Prüfung der Anmeldung beantragt. / Grant of a European patent, and examination of the application under Article 94, are hereby requested. / Il est demandé la délivrance d'un brevet européen et, conformément à l'article 94, l'examen de la demande. ☒ EXAM 4

Prüfungsantrag in einer zugelassenen Nichtamtssprache / Request for examination in an admissible non-EPO language / Requête en examen dans une langue non officielle autorisée

5.1 Der Anmelder verzichtet auf die Aufforderung nach Regel 70 (2), zu erklären, ob die Anmeldung aufrechterhalten wird. / The applicant waives his right to be asked whether he wishes to proceed further with the application (Rule 70(2)). / Le demandeur renonce à être invité, conformément à la règle 70(2), à déclarer s'il souhaite maintenir sa demande. ☐ MEPA

6 Zeichen des Anmelders oder Vertreters / Applicant's or representative's reference / Référence du demandeur ou du mandataire — AREF

Anmelder / Applicant / Demandeur — APPR

7 Name / Nom

8 Anschrift / Address / Adresse

9 Zustellanschrift / Address for correspondence / Adresse pour la correspondance

EPA/EPO/OEB 1001.1 – 10.19

TRAN | FILL

Zeichen des Anmelders / Applicant's reference / Référence du demandeur

1

Fig. 11.9 Application form European Patent—1 (EPO)

10 Staat des Wohnsitzes oder Sitzes /
State of residence or of principal place of business /
Etat du domicile ou du siège

11 Staatsangehörigkeit /
Nationality /
Nationalité

12 Telefon /
Telephone /
Téléphone

13 Fax /
Téléfax

14 Weitere(r) Anmelder auf Zusatzblatt / Additional applicant(s) on additional sheet /
Autre(s) demandeur(s) sur feuille supplémentaire

14.1 Der/Jeder Anmelder erklärt hiermit, eine Einheit oder eine natürliche Person nach Regel 6 (4) EPÜ zu sein. / The/Each applicant hereby declares that he is an entity or a natural person under Rule 6(4) EPC. /
Le/Chaque demandeur déclare par la présente être une entité ou une personne physique au sens de la règle 6(4) CBE

Vertreter / Representative / Mandataire FREP

15 Name / Nom
(**Nur einen** Vertreter oder den Namen des Zusammenschlusses angeben, der in das Europäische Patentregister einzutragen ist und an den zugestellt wird) /
(Name **only one** representative or association of representatives, to be listed in the Register of European Patents and to whom communications are to be notified) /
(N'indiquer qu'**un seul** mandataire ou le nom du groupement de mandataires qui sera inscrit au Registre européen des brevets et auquel les significations seront faites)

et al

16 Geschäftsanschrift /
Address of place of business /
Adresse professionnelle

17 Telefon /
Telephone /
Téléphone

18 Fax /
Téléfax

19 Weitere(r) Vertreter auf Zusatzblatt /
Additional representative(s) on additional sheet /
Autre(s) mandataire(s) sur feuille supplémentaire

Vollmacht / Authorisation / Pouvoir GENA

20 ist beigefügt / is enclosed / joint

21.1 Allgemeine Vollmacht ist registriert unter Nummer /
General authorisation has been registered under No. /
Un pouvoir général a été enregistré sous le numéro

21.2 Eine allgemeine Vollmacht wurde eingereicht, aber noch nicht registriert. /
A general authorisation has been filed but not yet registered. /
Un pouvoir général a été déposé mais n'est pas encore enregistré

Erfinder / Inventor / Inventeur INVT 20

22 Der (die) Anmelder ist (sind) alleinige(r) Erfinder. /
The applicant(s) is (are) the sole inventor(s). /
Le(s) demandeur(s) est (sont) le(s) seul(s) inventeur(s).

23 Erfindernennung in beigefügtem Schriftstück /
Designation of inventor attached /
Voir la désignation de l'inventeur ci-jointe

TIDE TIEN TIFR

24 **Bezeichnung der Erfindung / Title of invention / Titre de l'invention**

Zeichen des Anmelders /
Applicant's reference /
Référence du demandeur

EPA/EPO/OEB 1001.2 – 10.19

2

Fig. 11.10 Application form European Patent—2 (EPO)

45 Für Angestellte nach Artikel 133 (3) Satz 1 mit allgemeiner Vollmacht / For employees under Article 133(3), first sentence, having a general authorisation / Pour les employés mentionnés à l'article 133(3), première phrase, munis d'un pouvoir général

Nummer / Number / Numéro

46 **Unterschrift(en) des (der) Anmelder(s) oder Vertreter(s)**
Name des (der) Unterzeichneten bitte in Druckschrift wiederholen und bei juristischen Personen die Stellung des (der) Unterzeichneten innerhalb der Gesellschaft angeben. /
Signature(s) of applicant(s) or representative(s)
Under signature please print name and, in the case of legal persons, position within the company. /
Signature(s) du (des) demandeur(s) ou du (des) mandataire(s)
Prière d'indiquer en caractères d'imprimerie le ou les noms des signataires ainsi que, s'il s'agit d'une personne morale, la position occupée au sein de celle-ci par le ou les signataires.

Ort / Place / Lieu

Datum / Date

Unterschrift(en) / Signature(s)

EPA/EPO/OEB 1001.8 – 10.19

Zeichen des Anmelders / Applicant's reference / Référence du demandeur

8

Fig. 11.11 Application form European Patent—3 (EPO)

11.4 Application for an International Patent Application

In the application for filing an international patent application, the title of the application (designation of the invention) can first be entered in field No. I. In field No. II, the applicant's details are entered. In field No. X, the applicant's signature must be provided (see Figs. 11.12 and 11.13). The form can be accessed via the link "https://www.wipo.int/export/sites/www/pct/de/forms/request/ed_request.pdf" of WIPO.[9]

[9] World Intellectual Property Organization (WIPO), 34, chemin des Colombettes, CH-1211 Geneva 20, Switzerland.

PCT

ANTRAG

Der Unterzeichnete beantragt, daß die vorliegende internationale Anmeldung nach dem Vertrag über die internationale Zusammenarbeit auf dem Gebiet des Patentwesens behandelt wird.

Vom Anmeldeamt auszufüllen

Internationales Aktenzeichen

Internationales Anmeldedatum

Name des Anmeldeamts und "PCT International Application"

Aktenzeichen des Anmelders oder Anwalts *(falls gewünscht)* *(max. 25 Zeichen)*

Feld Nr. I BEZEICHNUNG DER ERFINDUNG

Feld Nr. II ANMELDER ☐ Diese Person ist gleichzeitig Erfinder

Name und Anschrift: *(Familienname, Vorname; bei juristischen Personen vollständige amtliche Bezeichnung. Bei der Anschrift sind die Postleitzahl und der Name des Staats anzugeben. Der in diesem Feld in der Anschrift angegebene Staat ist der Staat des Sitzes oder Wohnsitzes des Anmelders, sofern nachstehend kein Staat des Sitzes oder Wohnsitzes angegeben ist.)*

Telefonnr.:

Telefaxnr.:

Registrierungsnr. des Anmelders beim Amt:

E-Mail-Ermächtigung: Durch Ankreuzen eines der Kästchen werden das Anmeldeamt, die Internationale Recherchenbehörde, das Internationale Büro und die mit der internationalen vorläufigen Prüfung beauftragte Behörde ermächtigt, die in diesem Feld angegebene E-Mail-Adresse zu benutzen, um Mitteilungen bezüglich dieser internationalen Anmeldung zu übersenden, soweit das Amt oder die Behörde dazu bereit ist.

☐ nur für Vorauskopien, Mitteilungen werden zudem in Papierform versandt, oder

☐ ausschließlich in elektronischer Form (Mitteilungen werden nicht in Papierform versandt)

E-Mail-Adresse:

Staatsangehörigkeit *(Staat)*:

Sitz oder Wohnsitz *(Staat)*:

Diese Person ist Anmelder für folgende Staaten: ☐ alle Bestimmungsstaaten ☐ die im Zusatzfeld angegebenen Staaten

Feld Nr. III WEITERE ANMELDER UND/ODER (WEITERE) ERFINDER

☐ Weitere Anmelder und/oder (weitere) Erfinder sind auf einem Fortsetzungsblatt angegeben.

Feld Nr. IV ANWALT ODER GEMEINSAMER VERTRETER; ODER ZUSTELLANSCHRIFT

Die folgende Person wird hiermit bestellt/ist bestellt worden, um für den (die) Anmelder vor den zuständigen internationalen Behörden in folgender Eigenschaft zu handeln als: ☐ Anwalt ☐ gemeinsamer Vertreter

Name und Anschrift: *(Familienname, Vorname; bei juristischen Personen vollständige amtliche Bezeichnung. Bei der Anschrift sind die Postleitzahl und der Name des Staats anzugeben.)*

Telefonnr.:

Telefaxnr.:

Registrierungsnr. des Anwalts beim Amt:

E-Mail-Ermächtigung: Durch Ankreuzen eines der Kästchen werden das Anmeldeamt, die Internationale Recherchenbehörde, das Internationale Büro und die mit der internationalen vorläufigen Prüfung beauftragte Behörde ermächtigt, die in diesem Feld angegebene E-Mail-Adresse zu benutzen, um Mitteilungen bezüglich dieser internationalen Anmeldung zu übersenden, soweit das Amt oder die Behörde dazu bereit ist.

☐ nur für Vorauskopien, Mitteilungen werden zudem in Papierform versandt, oder

☐ ausschließlich in elektronischer Form (Mitteilungen werden nicht in Papierform versandt)

E-Mail-Adresse:

☐ **Zustellanschrift:** Dieses Kästchen ist anzukreuzen, wenn kein Anwalt oder gemeinsamer Vertreter bestellt ist und statt dessen im obigen Feld eine spezielle Zustellanschrift angegeben ist.

Formblatt PCT/RO/101 (Blatt 1) (Juli 2020)

Siehe Anmerkungen zu diesem Antragsformular

Fig. 11.12 Application form International Patent Application—1 (WIPO)

Blatt Nr. ....

Feld Nr. IX **KONTROLLISTE für in Papierform eingereichte Anmeldungen** - Dieses Blatt sollte nur für auf **Papier** eingereichte internationale Anmeldungen benutzt werden

Die internationale Anmeldung **enthält folgendes:**	Anzahl an Blättern
(a) Antragsformular PCT/RO/101 (inklusive eventueller Erklärungs- und Zusatzblätter)	:
(b) Beschreibung (ohne Sequenzprotokoll der Beschreibung, siehe unter (f))	:
(c) Ansprüche..................	:
(d) Zusammenfassung	:
(e) Zeichnungen (falls vorhanden)............	:
(f) Sequenzprotokoll der Beschreibung (falls vorhanden)............	:
Gesamtanzahl	: 0

Dieser internationalen Anmeldung liegen die folgenden Unterlagen bei *(kreuzen Sie die entsprechenden Kästchen an und geben Sie in der rechten Spalte jeweils die Anzahl der beiliegenden Exemplare an)* — Anzahl

1. ☐ Blatt für die Gebührenberechnung .... :
2. ☐ Original einer gesonderten Vollmacht.................. :
3. ☐ Original einer allgemeinen Vollmacht.................. :
4. ☐ Kopie der allgemeinen Vollmacht; Aktenzeichen: .... :
5. ☐ Prioritätsbeleg(e), in Feld Nr. VI durch folgende Zeilennummer(n) gekennzeichnet: .... :
6. ☐ Übersetzung der internationalen Anmeldung in die folgende Sprache: .... :
7. ☐ Gesonderte Angaben zu hinterlegten Mikroorganismen oder anderem biologischen Material .... :
8. ☐ *(nur wenn Punkt (f) in der linken Spalte markiert ist)* Kopie des Sequenzprotokolls in elektronischer Form (Anhang C/ST.25 Textdatei) auf einem physischen Datenträger, die nach Regel 13*ter* ausschließlich der internationalen Recherche dient und nicht Bestandteil der internationalen Anmeldung ist *(Art und Anzahl der physischen Datenträger)* .... :
9. ☐ *(nur wenn Punkt (f) (in der linken Spalte) und Punkt 8 (oben) markiert sind)* Erklärung, daß die nach Regel 13*ter* in elektronischer Form eingereichten Daten mit dem in Papierform eingereichten in der internationalen Anmeldung enthaltenen Sequenzprotokoll übereinstimmen........... :
10. ☐ Kopie der Ergebnisse von (einer) früheren Recherche(n) (Regel 12*bis*.1 Absatz a) .... :
11. ☐ Sonstige *(einzeln aufführen)*: .... :

Abbildung der Zeichnungen, die mit der Zusammenfassung veröffentlicht werden soll:

Sprache, in der die internationale Anmeldung eingereicht wird:

Feld Nr. X **UNTERSCHRIFT DES ANMELDERS, DES ANWALTS ODER DES GEMEINSAMEN VERTRETERS**

Der Name jeder unterzeichnenden Person ist neben der Unterschrift zu wiederholen, und es ist anzugeben, sofern sich dies nicht eindeutig aus dem Antrag ergibt, in welcher Eigenschaft die Person unterzeichnet.

Vom Anmeldeamt auszufüllen

1. Datum des tatsächlichen Eingangs dieser internationalen Anmeldung:
2. Zeichnungen: ☐ eingegangen: ☐ nicht eingegangen:
3. Geändertes Eingangsdatum aufgrund nachträglich, jedoch fristgerecht eingegangener Unterlagen oder Zeichnungen zur Vervollständigung dieser internationalen Anmeldung:
4. Datum des fristgerechten Eingangs der angeforderten Richtigstellungen nach Artikel 11(2) PCT:
5. Internationale Recherchenbehörde *(falls zwei oder mehr zuständig sind)*:
6. ☐ Übermittlung des Recherchenexemplars bis zur Zahlung der Recherchengebühr aufgeschoben

Vom Internationalen Büro auszufüllen

Datum des Eingangs des Aktenexemplars beim Internationalen Büro:

Formblatt PCT/RO/101 (letztes Blatt des Antragsformulars) (Juli 2020) *Siehe Anmerkungen zu diesem Antragsformular*

Fig. 11.13 Application form International Patent Application—2 (WIPO)

Search for Prior Art 12

A search of the state of the art should be conducted before drafting a patent or utility model application. This allows for the identification of which features are new and inventive in light of the prior art. The applicant should focus on the clear and precise description of these features.

Research can be conducted in particular in four online databases: the database of the German Patent Office (Depatisnet.de), the database of the European Patent Office (Espacenet), the database of the World Intellectual Property Organization WIPO (Patentscope), and the patent database of Google (Google Patents). All four databases can be accessed online and free of charge.

12.1 Depatisnet.de (German Patent and Trademark Office)

The online database Depatisnet.de of the German Patent and Trademark Office is the database used by examiners at the German Patent and Trademark Office to search for prior art documents when assessing the patentability of a patent application. It contains over 88 million patent specifications, published applications, examined applications, and utility model documents from all major industrialized nations worldwide.

The link "https://depatisnet.dpma.de/DepatisNet/depatisnet?action=einsteiger" leads to the beginner's search of the German Patent and Trademark Office's online database.[1] Here, in particular, the publication number of a patent document (first line of the input form) and/or a keyword under "Full-text search" (last line of the input form) can be entered (see Fig. 12.1).

[1] German Patent and Trademark Office (DPMA), Zweibrückenstrasse 12, 80331 Munich.

T. H. Meitinger, *Patent Made Simple: How to File Without a Lawyer*,
https://doi.org/10.1007/978-3-662-73096-6_12

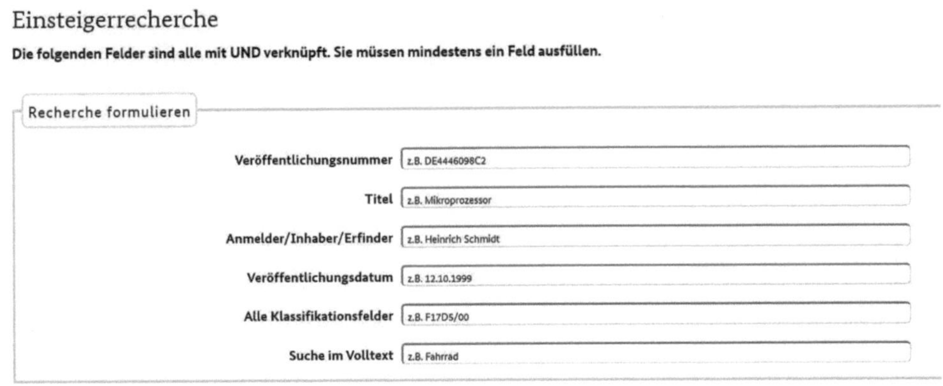

Fig. 12.1 Depatisnet.de entry form (DPMA)

12.2 Espacenet (EPO, European Patent Office)

The Espacenet web portal provides access to the EPODOC database of the European Patent Office. EPODOC contains over 120 million patent documents and can be accessed via the link "https://worldwide.espacenet.com/patent/my-espacenet" of the European Patent Office[2] (see Fig. 12.2).

[2] European Patent Office (EPO), Bob-van-Benthem-Platz 1, 80469 Munich.

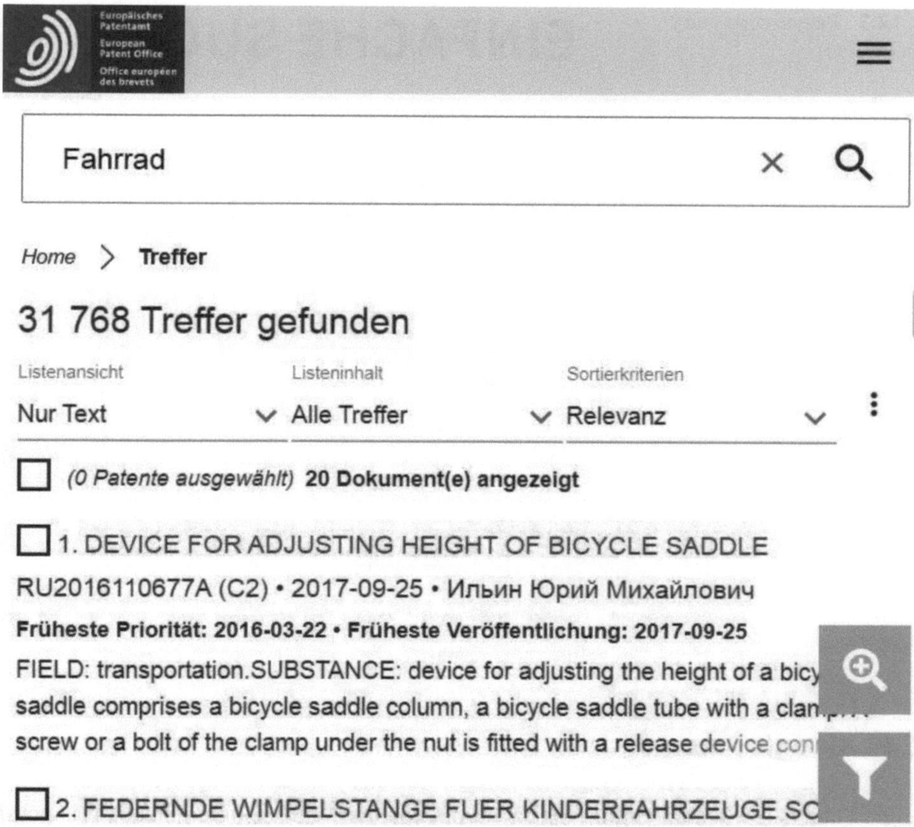

Fig. 12.2 Espacenet entry form (EPO)

12.3 Patentscope (WIPO, World Intellectual Property Organization)

With the WIPO Patentscope online database, it is possible to search over 95 million patent documents. PatentScope can be accessed via the link "https://patentscope.wipo.int/search/de/search.jsf" of WIPO[3] (see Fig. 12.3). In the "Field" input option, for example, "Full text" can be selected to search the complete patent documents for a keyword ("search term").

[3] World Intellectual Property Organization, 34, chemin des Colombettes, CH-1211 Geneva 20, Switzerland.

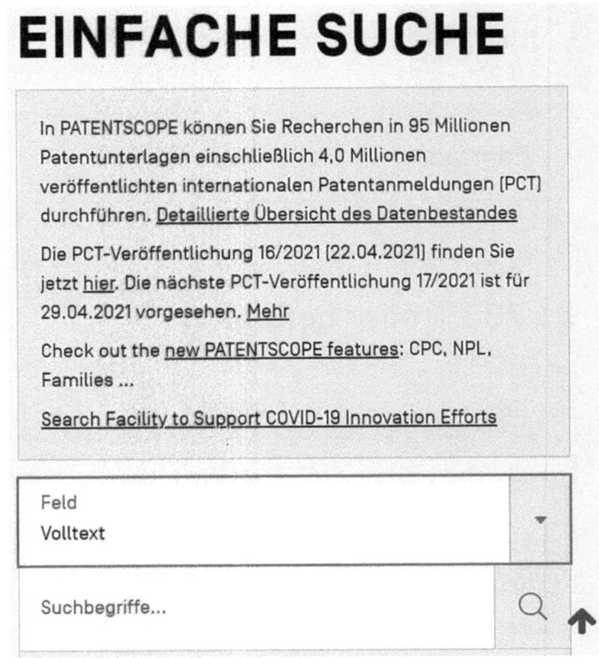

Fig. 12.3 Patentscope entry form (WIPO)

12.4 Google Patents

Google Patents is a search engine for patent documents. With Google Patents, it is possible to search the patent records of 17 patent offices. In particular, Google Patents allows searching of the patent documents of the US Patent and Trademark Office USPTO,[4] the European Patent Office EPO, the Japan Patent Office JPO,[5] the China National Intellectual Property Administration CNIPA,[6] the World Intellectual Property Organization WIPO, and the German Patent and Trademark Office DPMA. An advantage is that, with Google Translate, patent information can be directly translated into the user's language. The link "https://patents.google.com/" leads to the input form of Google Patents[7] (see Fig. 12.4).

[4] USPTO: United States Patent and Trademark Office.

[5] JPO: Japan Patent Office.

[6] CNIPA: China National Intellectual Property Administration.

[7] Google Ireland Limited, Gordon House, Barrow Street, Dublin 4, Ireland.

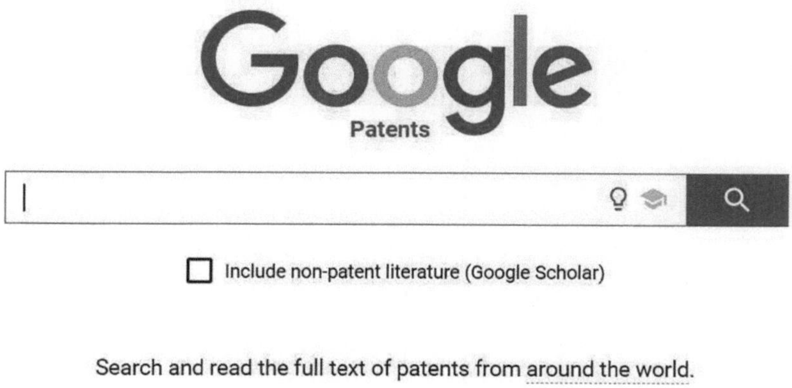

Fig. 12.4 Google Patents entry form (Google LLC)

Employee Inventions 13

For the inventive employee, the Act on Employee Inventions applies.[1] According to the Employee Inventions Act, there are three types of inventions: service inventions, released inventions, and free inventions.

13.1 Service Invention and Released Invention

Service inventions are made during an employment relationship and arise from the employee's work activities or are substantially based on the experience of the company.[2] A service invention must be reported to the employer without delay, separately and in written form. It must be made clear that this is a notification of an invention.[3] The service invention can be claimed by the employer by declaration to the employee.[4] Furthermore, a claim is deemed to have been made if the employer does not release the service invention in written form within four months after receipt of the invention disclosure.[5] Upon claiming, the rights to the invention transfer to the employer.[6] Otherwise, the

[1] Act on Employee Inventions in the version published in the Federal Law Gazette Part III, classification number 422–1, as last amended by Article 7 of the Act of July 31, 2009 (BGBl. I p. 2521).
[2] Section 4 paragraph 2 Employee Inventions Act.
[3] Section 5 paragraph 1 sentence 1 Employee Inventions Act.
[4] Section 6 paragraph 1 Employee Inventions Act.
[5] Section 6 paragraph 2 Employee Inventions Act.
[6] Section 7 paragraph 1 Employee Inventions Act.

T. H. Meitinger, *Patent Made Simple: How to File Without a Lawyer*,
https://doi.org/10.1007/978-3-662-73096-6_13

employee may freely dispose of the released invention and may also file for a patent or utility model in their own name.[7]

13.2 Free Invention

An invention that is not a service invention is a free invention.[8] A free invention must be reported to the employer so that the employer can verify that it is indeed a free invention, i.e., an invention that is unrelated to the employee's work activities or based on the company's know-how.[9] If the employer does not contest within three months of receiving the employee's notification that it is a free invention, claiming the invention as a service invention is excluded.[10] If the invention is obviously not usable by the employer, there is no obligation to report the free invention.[11] As a rule, however, one should still report the invention to the employer to be on the safe side. It is actually very difficult to assess whether an invention is not in fact a service invention or whether it is truly not usable by the employer. Once the three months have passed, the invention can be filed for a patent or utility model in one's own name without legal risk.

The employee has a duty to offer the invention to the employer. A free invention must first be offered to the employer during the term of employment before it is exploited elsewhere. At a minimum, the employer must be offered a non-exclusive right to use the invention on reasonable terms. The prerequisite is that, at the time of the offer, the invention falls within the existing or prepared field of activity of the employer's business.[12] The employer has three months to accept the offer. After that, the right of first refusal expires.[13]

[7] Section 8 sentence 2 Employee Inventions Act.
[8] Section 4 paragraph 3 sentence 1 Employee Inventions Act.
[9] Section 18 paragraph 1 Employee Inventions Act.
[10] Section 18 paragraph 2 Employee Inventions Act.
[11] Section 18 paragraph 3 Employee Inventions Act.
[12] Section 19 paragraph 1 sentence 1 Employee Inventions Act.
[13] Section 19 paragraph 2 Employee Inventions Act.

14 Joint Inventorship

An invention can be created by a single inventor. However, an invention can also originate from several inventors. A group of inventors holds the right to the invention jointly. In the case of a group of inventors, the question arises as to the extent to which an individual inventor may exploit the invention without regard to the other inventors.

If no alternative arrangement has been made, a group of inventors constitutes what is known as a fractional co-ownership under §§ 741 et seq. BGB.[1] Each inventor may use the invention independently for their own purposes.[2] For example, if several companies have jointly filed an invention, each company can offer products and services based on the patent for its own market segments.

A particular situation may arise if a natural person, without the means to commercially exploit the invention, and a company have jointly developed the invention and filed for a patent. In this case, it may be that the natural person is unable to exploit the invention, whereas the company can manufacture innovative products based on the invention. For the natural person, the only way to realize value from the invention may be to secure a competitor of the company as a licensee. Naturally, this can create potential for conflict within the group of inventors or applicants. The company could, in particular under § 747 sentence 2 BGB, prevent licensing to a competitor. Such a situation can result in the natural person being entirely unable to exploit the invention. The only remaining value from the creation of the invention for the natural person is then the designation as inventor.[3]

[1] BGH, X ZR 223/98, 'Rollenantriebseinheit', *Industrial Property Protection and Copyright Law,* GRUR, 2001, 226.

[2] § 743 paragraph 2 BGB

[3] § 63 paragraph 1 sentence 1 Patent Act.

T. H. Meitinger, *Patent Made Simple: How to File Without a Lawyer*,
https://doi.org/10.1007/978-3-662-73096-6_14

Compensation claims could be asserted so that the natural person would be entitled to compensation payments from the company.[4] This could achieve an appropriate balance of interests. The Federal Court of Justice, however, did not adopt this view. The Federal Court of Justice follows the principle that the diligent and industrious inventor should not be penalized in order to reward the idle inventor. For the Federal Court of Justice, there is no room for compensation claims.[5]

[4] Ernst Fischer, 'Exploitation Rights in Patent Co-ownership', *Industrial Property Protection and Copyright Law,* GRUR, 1977, 313–318.

[5] BGH, X ZR 152/03, 'Gummielastische Masse II', *Industrial Property Protection and Copyright Law,* GRUR, 2005, 663–665.

15 Utility Model

The preparation of a utility model application essentially corresponds to the drafting of a patent application. The differences are that, in the case of a utility model, the term "protection claims" is used instead of "patent claims", and a utility model application does not include an abstract. Otherwise, the invention in a utility model application must also be described comprehensively, clearly, and concisely.[1] However, in a utility model, the inventor does not need to be named.

A utility model application must first explain the prior art and discuss its disadvantages. These disadvantages give rise to the technical problem, which is then followed by a detailed description of the invention. Various examples and embodiments should be explained. To illustrate the invention, it is advisable to use one or more drawings. In addition, protection claims must be attached to the documents. The essential features of the invention must be set out in these protection claims. These protection claims define what is being claimed as intellectual property.

The Utility Model Act does not provide for an official examination of the legal validity of an invention. An examination of novelty or inventive step by the Patent Office is excluded. At most, an official search can be requested.[2] Accordingly, the Patent Office can only issue notifications regarding a utility model application if there are formal deficiencies.

---

[1] Section 34, paragraph 4 of the Patent Act: "The invention must be disclosed in the application in such a clear and complete manner that a person skilled in the art can carry it out." Analogous: Article 83 EPC. The applicant should explain the invention comprehensively and disclose all embodiments. This avoids deficiencies in clarity or feasibility and enables meaningful fallback positions to distinguish from the prior art in cancellation proceedings.

[2] Section 7, paragraph 1 of the Utility Model Act.

T. H. Meitinger, *Patent Made Simple: How to File Without a Lawyer*,
https://doi.org/10.1007/978-3-662-73096-6_15

Formal deficiencies usually concern the drawings. The exact requirements for correct drawings can be found in the Utility Model Ordinance.[3] Otherwise, an official notification may be issued if a method[4] is claimed in the set of claims, or if the main claim or a dependent claim contains only non-technical features.

15.1 Little Brother of the Patent

The utility model is often referred to as the little brother of the patent or as a "small" patent. This carries the assumption that it is easier to obtain a utility model or that overcoming the inventive step, which is required for a legally valid utility model, is easier than achieving inventive step for a patent. This is a misconception. In fact, the same inventive level is required for the legal validity of a utility model as for a patent.[5]

15.2 Grace Period for Novelty

Utility model law provides for a general grace period for novelty of six months. Accordingly, disclosures or uses that occurred within six months prior to the filing date of the utility model application are not considered relevant prior art, provided they are based on the applicant's or their legal predecessor's work.[6] If, for example, the invention was previously published on the applicant's own website, a legally valid utility model can still be filed. In this case, patent grant is excluded due to lack of novelty.

The grace period for novelty should be regarded as a last resort. Under no circumstances should the invention be published before filing the utility model application in reliance on the grace period. If the publication spreads, it may be difficult to prove that the original disclosure of the invention came from the applicant.

15.3 Limited Prior Art

Another advantage of the utility model is that, when assessing its legal validity, a narrower scope of prior art is considered compared to a patent. The relevant prior art includes all knowledge made available to the public by written description or by use in

---

[3] Section 7 of the Utility Model Ordinance.

[4] According to Section 2 No. 3 of the Utility Model Act, a utility model cannot claim a method.

[5] Federal Court of Justice, X ZB 27/05, 'Demonstration Cabinet', *Neue Juristische Wochenschrift,* 2006, 3208.

[6] Section 3, paragraph 1, sentence 3 of the Utility Model Act.

Table 15.1 Fees of the DPMA in Utility Model Matters

Utility Model Matters	
Filing fee (paper form—not online)	40 €
Search request	250 €
Maintenance fee for 4th to 6th year of protection	210 €
Maintenance fee for 7th and 8th year of protection	350 €
Maintenance fee for 9th and 10th year of protection	530 €

Germany.[7] An oral description of the invention therefore does not count as prior art for a utility model. For example, if the invention is explained to a fellow passenger on a train journey, this does not constitute relevant prior art. In addition, public use of the invention abroad is also irrelevant.

15.4 Fees for a Utility Model

A filing fee is charged for a utility model, which amounts to 40 euros for paper submissions. In addition, maintenance fees must be paid for the 4th to 6th year of protection (210 euros), the 7th and 8th year of protection (350 euros), and the 9th and 10th year of protection (530 euros). If a search request is filed, a fee of 250 euros is payable (see Table 15.1).[8]

[7] Section 3, paragraph 1, sentence 2 of the Utility Model Act.

[8] Annex to Section 2, paragraph 1 of the Patent Costs Act (Schedule of Fees, as of April 25, 2026).

German Patent

16

An application can be filed with the German Patent and Trademark Office either as a utility model application or as a patent application. Alternatively, a patent application can be filed with the European Patent Office as a European patent application. There is also the option to pursue an international application. The application documents for an international patent application can be submitted, at the applicant's choice, to the German Patent and Trademark Office (DPMA), the European Patent Office (EPO), or directly to the World Intellectual Property Organization (WIPO).

16.1 Examination or Search?

A German patent application can be filed with the Patent Office without the grant procedure commencing immediately. The applicant has seven years to file a request for examination.[1] Alternatively, the applicant may first file a request for search.[2] If the applicant does not file a request for examination at the same time as or shortly after submitting the application documents, the Patent Office will not assign a high priority to the examination request. Rapid processing, and in particular a first official communication within the priority period, cannot then be expected.

[1] Section 44, paragraph 2, sentence 1 of the Patent Act.

[2] Section 43, paragraph 1, sentence 1 of the Patent Act.

T. H. Meitinger, *Patent Made Simple: How to File Without a Lawyer*,
https://doi.org/10.1007/978-3-662-73096-6_16

Table 16.1 DPMA Fees in Patent Matters

Patent Matters	
Filing fee (paper form—not online)	€60
Search request	€300
Examination request	€350

16.2 Fees for a German Patent

The costs of a patent application can be divided into two categories: official fees and patent attorney fees. Official fees include the filing fee, the fee for a search or examination request, and the annual fees for maintaining the intellectual property right. Patent attorney fees vary widely. Unlike attorneys-at-law, there is no fixed fee schedule for patent attorneys. Depending on the complexity, the preparation of a patent or utility model application can cost between 2,000 and 6,000 euros.

For filing a patent application in paper form, a filing fee of €60 must be paid to the Patent Office. The fee for a search request is €300, and for an examination request, €350 must be paid (see Table 16.1).[3]

For the third patent year and each subsequent year, an annual fee must be paid to the Patent Office. Initially, the annual fee is relatively low but increases progressively. For the first ten years of a patent, the total annual fees amount to €1.663. For the full term of 20 years, the total cost of annual fees is €14.163 (see Table 16.2).[4]

[3] Annex to Section 2, paragraph 1 of the Patent Costs Act (Schedule of Fees), as of April 25, 2026.

[4] Annex to Section 2, paragraph 1 of the Patent Costs Act (Schedule of Fees), as of April 25, 2026.

Table 16.2 Annual Fees of the DPMA

Maintenance of a Patent or Application	
Annual fee for the 3rd patent year	€70
Annual fee for the 4th patent year	€70
Annual fee for the 5th patent year	€100
Annual fee for the 6th patent year	€150
Annual fee for the 7th patent year	€210
Annual fee for the 8th patent year	€280
Annual fee for the 9th patent year	€350
Annual fee for the 10th patent year	€430
Annual fee for the 11th patent year	€540
Annual fee for the 12th patent year	€680
Annual fee for the 13th patent year	€830
Annual fee for the 14th patent year	€980
Annual fee for the 15th patent year	€1,130
Annual fee for the 16th patent year	€1,310
Annual fee for the 17th patent year	€1,490
Annual fee for the 18th patent year	€1,670
Annual fee for the 19th patent year	€1,840
Annual fee for the 20th patent year	2.030

European Patent 17

With a European application, a patent can be obtained in all member states of the European Patent Convention (EPC). For this purpose, a unified grant procedure is provided. The member states of the EPC are not the same as those of the EU (European Union).

17.1 Member States of the European Patent Convention

The member states of the EPC include, in particular, Germany, the United Kingdom, France, Italy, Spain, and the Netherlands. Table 17.1 lists the member states in order of accession date (the first column contains the country codes, the third column shows the accession date to the EPC) (see Table 17.1).[1]

[1] As of: 7 May 2021.

T. H. Meitinger, *Patent Made Simple: How to File Without a Lawyer*,
https://doi.org/10.1007/978-3-662-73096-6_17

Table 17.1 Member States of the EPC

BE	Belgium	7 October 1977
DE	Germany	7 October 1977
FR	France	7 October 1977
LU	Luxembourg	7 October 1977
NL	Netherlands	7 October 1977
CH	Switzerland	7 October 1977
GB	United Kingdom	7 October 1977
SE	Sweden	1 May 1978
IT	Italy	1 December 1978
AT	Austria	1 May 1979
LI	Liechtenstein	1 April 1980
GR	Greece	1 October 1986
ES	Spain	1 October 1986
DK	Denmark	1 January 1990
MC	Monaco	1 December 1991
PT	Portugal	1 January 1992
IE	Ireland	1 August 1992
FI	Finland	1 March 1996
CY	Cyprus	1 April 1998
TR	Turkey	1 November 2000
BG	Bulgaria	1 July 2002
CZ	Czech Republic	1 July 2002
EE	Estonia	1 July 2002
SK	Slovakia	1 July 2002
SI	Slovenia	1 December 2002
HU	Hungary	1 January 2003
RO	Romania	1 March 2003
PL	Poland	1 March 2004
IS	Iceland	1 November 2004
LT	Lithuania	1 December 2004
LV	Latvia	1 July 2005
MT	Malta	1 March 2007
HR	Croatia	1 January 2008
NO	Norway	1 January 2008
MK	North Macedonia	1 January 2009
SM	San Marino	1 July 2009
AL	Albania	1 May 2010
RS	Serbia	1 October 2010

17.2 Search

When submitting the application documents to the European Patent Office, the applicant must pay both the filing fee and the search fee.[2] The fees must be paid within one month from the filing date.[3] If the fees are not paid within the prescribed period, the application is deemed to be withdrawn.[4] Alternatively, the applicant may, when filing the application, request examination of the application. In this case, the Search Division is bypassed and the grant procedure is initiated directly.

17.3 Examination

After publication[5] of the search report, the applicant is given a period of six months to pay the examination fee and thereby initiate the examination procedure. Otherwise, the application is deemed to be withdrawn.[6]

17.4 Fees for a European Patent

For a European patent application, a filing fee of €285 must be paid. In addition, a search fee of €1.595 is due upon submission of the application documents. If a request for examination is filed after receipt of the search report, a fee of €2.010 is payable. At the same time as the examination fee, a designation fee of €720 must be paid (see Table 17.2).[7]

In addition, from the third patent year and for each subsequent year, an annual fee must be paid to maintain the European patent application (see Table 17.3).[8]

[2] Article 78(2), first sentence, EPC.

[3] Rule 38(1) EPC.

[4] Article 78(2), second sentence, EPC.

[5] More precisely: the six-month period begins after the notice of publication of the European search report in the European Patent Bulletin (Rule 70(1), first sentence EPC).

[6] Rule 70(3) EPC.

[7] EPO, https://fees.apps.epo.org/prod/ui/?lang=DE&page=1&currency=EUR, accessed on 25 April 2026.

[8] EPO, https://fees.apps.epo.org/prod/ui/?lang=DE&page=1&currency=EUR, accessed on 25 April 2026.

Table 17.2 Fees of the EPO

Patent matters	
Filing fee (paper form—not online)	€285
Search fee	€1.595
Examination fee	€2.010

Table 17.3 Annual fees of the EPO

Maintenance of a patent or application	
Annual fee for the 3rd patent year	€725
Annual fee for the 4th patent year	€885
Annual fee for the 5th patent year	1.050
Annual fee for the 6th patent year	€1.215
Annual fee for the 7th patent year	€1.375
Annual fee for the 8th patent year	€1.540
Annual fee for the 9th patent year	€1.700
Annual fee for the 10th and each subsequent patent year	€1.865

18 International Patent Application

With an international application,[1] the priority year can effectively be extended. This means that a national grant procedure in almost any country in the world[2] can be initiated 30 or 31 months after the filing date or priority date, with the priority of the international application.[3] Therefore, an international application does not result in a patent, but merely postpones the start of the respective national examination procedure into the future. The application documents for an international patent application can be filed directly with WIPO in Geneva, with the German Patent and Trademark Office (DPMA), or with the European Patent Office (EPO). It is advisable to file an international application with the European Patent Office, as the European Patent Office is the ISA (International Search Authority).

18.1 Search Report and Written Opinion

Together with the submission of the application documents, a request for search must be filed and a search fee paid.[4] A search report is sent to the applicant together with a written opinion,[5] from which the search authority's assessment of patentability can be derived.

---

[1] Article 3(1) PCT.

[2] Important exceptions are Argentina, Taiwan, and Venezuela.

[3] In some countries, the national grant procedure must be started no later than 30 months (examples: China, Canada, and the USA), in other countries no later than 31 months (examples: South Korea and Russia).

[4] Article 15(1) PCT.

[5] Rule 43*bis* PCT.

T. H. Meitinger, *Patent Made Simple: How to File Without a Lawyer*,
https://doi.org/10.1007/978-3-662-73096-6_18

Table 18.1 Fees for an International Patent Application

International Patent Application	
Filing and transmittal fee (EPO)	€1.578
Search fee	€1.845
Examination fee	€2.010

18.2 International Preliminary Examination Report

In the international procedure, it is possible to file a request for examination.[6] As a rule, no request for examination is filed, since the so-called international preliminary examination report usually corresponds to the written opinion that has already been sent to the applicant together with the international search report. Furthermore, the international preliminary examination report is not binding for a subsequent regional or national patent office.

18.3 Fees for an International Application

When filing an international patent application with the European Patent Office,[7] a filing and transmittal fee of €1.578 and a search fee of €1.845 must be paid (see Table 18.1).[8]

[6] Article 31(1) PCT.

[7] It is advisable to file an international application with the European Patent Office, as the European Patent Office is the International Search Authority. (Article 16(1) PCT)

[8] WIPO, PCT Fee Tables, https://www.wipo.int/export/sites/www/pct/en/fees.pdf and EPO, https://my.epoline.org/epoline-portal/classic/epoline.Scheduleoffees, accessed on April 25, 2026).

Responding to an Official Action

19

In German patent law, the commencement of the grant procedure can be delayed for up to seven years.[1] Upon request, the examination procedure is initiated before the German Patent and Trademark Office. In the European procedure, a search for relevant prior art documents begins immediately after the filing of the application documents.

Deficiencies in the application, particularly in the claims, are communicated to the applicant in an official notice from the patent office. This notice must be "answered".[2] The applicant examines whether they can understand the examiner's assessment. If so, the claims and the description may need to be amended. Otherwise, it must be justified why, from the applicant's perspective, the claims are allowable.

Typically, a single response to an office action is not sufficient. Rather, a correspondence with the patent office ensues, with two to three office actions to be expected before the examiner makes a final decision. The decision results in either the grant of a patent based on the current patent claims or the rejection of the application.[3] This decision can be appealed before the Federal Patent Court in Munich.

---

[1] Section 44, paragraph 2, sentence 1 of the Patent Act.

[2] Section 45, paragraph 1, sentence 1 of the Patent Act or Article 94, paragraph 3 EPC.

[3] Section 48, sentence 1 of the Patent Act.

T. H. Meitinger, *Patent Made Simple: How to File Without a Lawyer*,
https://doi.org/10.1007/978-3-662-73096-6_19

19.1 Formal Deficiencies

Formal deficiencies usually concern the drawings. The exact requirements for correct drawings can be found in the Patent Ordinance.[4]

19.2 Lack of Clarity

An office action from the German or European Patent Office typically identifies three types of deficiencies: lack of clarity, lack of novelty, or lack of inventive step in the claims, particularly the main claim and the dependent claims. The objection of lack of clarity is controversial in German patent law, in contrast to European patent law. In European patent law, Article 84, sentence 2 EPC stipulates that the patent claims must be drafted clearly. This establishes the requirement of clarity as a condition for patentability. In the German Patent Act, there is no analogous provision corresponding to Article 84, sentence 2 EPC. There is, however, Section 34, paragraph 4, which states that the invention must be disclosed in the application in such a clear and complete manner that a person skilled in the art can carry it out. However, this only means that the entire application must enable the invention to be carried out, and not that the claims themselves must already be clearly and unambiguously formulated.

Section 34, paragraph 3, no. 3 of the Patent Act stipulates that patent claims must specify what is to be protected. From this provision, the German Patent and Trademark Office derives the requirement that claims must meet the clarity requirement. However, this view was not shared by the Federal Patent Court in a recent case.[5] Currently, it is therefore assumed that lack of clarity of the claims is not a ground for rejection in a German granting procedure.

19.3 Impermissible Extension

An impermissible extension exists if the description or the claims of the application go beyond what was described in the originally filed application documents.[6]

[4] Annex 2 to Section 12 of the Patent Ordinance (Standards for the submission of drawings).

[5] Federal Patent Court, 15 W (pat) 9/13, judgment of 24 June 2015—"Polyurethane foam".

[6] Section 38, sentence 1 of the Patent Act or Article 123, paragraphs 2 and 3 EPC. In addition, subsequently published patent documents must be taken into account when assessing novelty (Section 3, paragraph 2, sentence 1 of the Patent Act or Article 54, paragraph 3 EPC).

19.4 Lack of Novelty

Novelty is lacking if the invention forms part of the prior art. Prior art is everything that, before the filing or priority date of the patent application, has been made available to the public in written or oral form, by use, or in any other way.[7] If, therefore, all features of the main claim are fully disclosed in a single prior art document, the subject matter of the main claim is not novel. Suitable features from the description or the dependent claims must then be incorporated into the main claim to establish its novelty. These features must be suitable for solving the technical problem. The technical problem may need to be reformulated.

19.5 Lack of Inventive Step

An invention does not involve an inventive step if it is obvious to a person skilled in the art from the prior art.[8] If two or three prior art documents can be combined to arrive at the subject matter of the main claim, then the main claim is not inventive. However, the person skilled in the art must also have a reason to combine the relevant prior art documents. On the other hand, if a document advises the skilled person against combining the documents, the skilled person would not combine the technical teachings of these documents, and the main claim would be based on an inventive step.

19.6 Problem-Solution Approach

A suitable way to demonstrate in a response to an office action that a main claim is based on an inventive step is the application of the problem-solution approach.

The problem-solution approach was developed by the European Patent Office to exclude misleading ex post facto analysis. In hindsight (ex post), many inventions appear obvious, perhaps even trivial. Nevertheless, at the time the invention was created, it required inventive activity to arrive at the invention.[9]

The problem-solution approach comprises three sections, namely the determination of the closest prior art, the definition of the objective technical problem, and the assessment

[7] Section 3, paragraph 1 of the Patent Act or Article 54, paragraphs 1 and 2 EPC.

[8] Section 4, sentence 1 of the Patent Act or Article 56, sentence 1 EPC.

[9] EPO, Guidelines for Examination, Part G Chapter VII 8. Ex post facto analysis, https://www.epo.org/law-practice/legal-texts/html/guidelines/d/g_vii_8.htm, accessed on April 6, 2021.

of whether the invention is inventive in view of the prior art and the objective technical problem.[10]

Closest Prior Art: The closest prior art is a document, in particular a patent specification, a patent application, or a utility model, which represents the most promising starting point for the development of the invention to be assessed. Typically, it is a document whose technical teaching requires the fewest structural and functional modifications to arrive at the invention.[11]

Objective Technical Problem: The technical problem of the invention is to arrive at the invention by modifying the closest prior art. The technical problem thus defined is referred to as the "objective technical problem". The procedure here is to identify the differences between the invention and the closest prior art in terms of structure or function. These differences result in an effect produced by the invention. The objective technical problem is then to achieve this effect.[12]

Could/would approach: It is not sufficient to find a document which, in combination with the closest prior art, leads to the invention. If the skilled person would not consult this document to solve the inventive problem, it would be incorrect to retrospectively question the inventive step on the basis of this document. A document originating from a completely different technical field, or one that advises against combining its technical teaching with that of the closest prior art, would not have been used by the skilled person to solve the objective technical problem. It is therefore not enough that documents could be combined (could) to arrive at the invention. Rather, the skilled person must also have had a reason to combine these documents (would). If no documents can be found that bridge the gap from the closest prior art to the invention, the invention is based on an inventive step.[13]

---

[10] EPO: Guidelines for Examination, Part G Chapter VII 5. Problem-solution approach, https://www.epo.org/law-practice/legal-texts/html/guidelines/d/g_vii_5.htm, accessed on April 6, 2021.

[11] EPO: Guidelines for Examination, Part G Chapter VII 5.1 https://www.epo.org/law-practice/legal-texts/html/guidelines/d/g_vii_5_1.htm, accessed on April 6, 2021.

[12] EPO, Guidelines for Examination, Part G Chapter VII 5.2 https://www.epo.org/law-practice/legal-texts/html/guidelines/d/g_vii_5_2.htm, accessed on April 6, 2021.

[13] EPO, T 2/83, https://www.epo.org/law-practice/case-law-appeals/recent/t830002ep1.html, accessed on May 26, 2021; EPO, T 257/98, https://www.epo.org/law-practice/case-law-appeals/recent/t980257eu1.html, accessed on May 26, 2021; EPO, T 35/04, https://www.epo.org/law-practice/case-law-appeals/recent/t040035eu1.html, accessed on May 26, 2021.

Template for a Patent Application

20

A template for a patent application is presented. Notes in parentheses explain the template. Example entries are set in quotation marks and should be adapted to the specific case.

Template for a Patent or Utility Model Application (for a utility model application, the summary is omitted):
Our reference: "G0002/TM"

(Name and address of the applicant)

(Title of the application) "Tool coupling"

<u>DESCRIPTION</u>
FIELD OF THE INVENTION
The present invention relates to a device and a method for … .

BACKGROUND OF THE INVENTION
It is known in the prior art that … .

SUMMARY OF THE INVENTION
The prior art has the disadvantage that … .

An object, therefore, is to provide a device and a method such that …

The object is achieved by the features of the independent claim(s). Advantageous developments of the invention are described in the dependent claims.

T. H. Meitinger, *Patent Made Simple: How to File Without a Lawyer*,
https://doi.org/10.1007/978-3-662-73096-6_20

(The so-called support for the claims follows next, in which the claims are reproduced verbatim.)

As a first aspect of the invention, a (main claim) is provided.

As a second aspect of the invention, a (second independent claim) is provided.

Exemplary embodiments are described in the dependent claims.

According to an exemplary embodiment of the invention, a (dependent claim 1) is provided.

According to a further exemplary embodiment of the present invention, a (dependent claim 2) is provided.

In a further embodiment according to the invention, a (dependent claim 3) is provided.

According to an exemplary embodiment of the invention, a (dependent claim 4) is provided.

An idea of the invention can be seen as … (The invention can be described here again in other words. The essential inventive points and special embodiments should be addressed here).

The individual features can, of course, also be combined with one another, which may also result in advantageous effects that go beyond the sum of the individual effects.

BRIEF DESCRIPTION OF THE DRAWINGS

Further details and advantages of the invention will become apparent from the exemplary embodiments shown in the drawings. The drawings show:

Fig. 1 a …,
Fig. 2 a … and
Fig. 3 a … .

DETAILED DESCRIPTION OF EXEMPLARY EMBODIMENTS

Fig. 1 shows a … .
Fig. 2 shows a … .
Fig. 3 shows a … .

It should be noted that the term "comprising" does not exclude additional elements or method steps, just as the terms "a" and "an" do not exclude a plurality of elements or steps.

The reference numerals used serve solely to enhance understanding and should in no way be considered limiting, with the scope of protection of the invention being defined by the claims.

LIST OF REFERENCE NUMERALS

1. "Car"
2. "Engine"
3. "Steering wheel"

CLAIMS

1. “Car for transporting a person”, wherein …
2. “Car” according to claim 1, wherein …
3. “Car” according to one of claims 1 or 2, wherein …
4. “Car” according to one of the preceding claims, wherein …
5. “Car” according to one of the preceding claims, wherein …
6. Method for “manufacturing a car” according to one of the preceding claims, comprising the steps:
   “Painting” a … for …
7. Method according to claim 6, further comprising the step:
   “Bending” a … for …
8. Method according to one of claims 6 or 7, further comprising the step: …
9. Method according to one of claims 6 to 8, further comprising the step: …
10. Method according to one of claims 6 to 9, further comprising the step: …

SUMMARY

(Title of the application) “Tool coupling”
(Wording of the main claim)
[Fig. 1]

21 Template for a Response to an Official Action

A template for a response to an office action is presented. Notes in parentheses explain the template. Example entries are set in quotation marks and must be adapted to the specific case.

Template for a response to an office action for submission to the German or European Patent Office:

(German procedure:) In response to the office action dated “15 July 2021”:

(European procedure:) In response to the communication dated “15 July 2021”:

<u>**New Documents**</u>

New claims 1 to 10 are submitted, which replace the original set of claims.

<u>**Disclosure of the Claims**</u>

The subject matter of the new claims is disclosed at the following locations in the original documents:

<u>Claim 1:</u>

The new claim 1 comprises the following features:

- original claim 1, the new claim 1 comprises all features of the original claim 1;
- original claim “5”, the new claim 1 comprises all features of the original claim “5”;
- original claim “6”, the new claim 1 comprises all features of the original claim “6”;
- (new feature): the description of the originally filed application contains on “page 12, second paragraph” the feature that …

T. H. Meitinger, *Patent Made Simple: How to File Without a Lawyer*,
https://doi.org/10.1007/978-3-662-73096-6_21

Claim “3”:

The new claim “3” comprises the following features:

- original claim “3”, the new claim “3” comprises all features of the original claim “3”;
- original claim “9”, the new claim “3” comprises all features of the original claim “9”;
- (An additional feature can be “incorporated” from the description.) “new feature”: the description of the originally filed application contains on “page 12, second paragraph” the feature that ...

The further claims of the new set of claims correspond to the previous set of claims.

Thus, the subject matter of the new claims does not extend beyond the content of the original application documents. The dependencies have been adjusted accordingly.

Novelty

Claim 1

D1 (DE...) describes a device for “applying paints or impregnation agents.” Document D2 (DE ...) discloses a device for “gluing chipboard elements or for introducing glue into grooves and dowel holes”. In contrast, the current claim 1 describes that ...

Therefore, the subject matter of claim 1 is novel, and consequently so are the subject matters of dependent claims 2 to 10.

Inventive Step

Claim 1

Document D1 represents the closest prior art, as D1 describes a device for ...

In contrast to the main claim, D1 does not disclose that ...

This feature has the effect/result that ...

The objective technical problem is therefore to provide a device that can achieve this effect/result, namely ...

This problem is solved according to the invention by the features of the current claim 1 in that ...

The subject matter of claim 1 is not obvious in view of D1 or D2, or in view of D1 in combination with common general knowledge, or in view of D2 and common general knowledge, or in view of a combination of D1 with D2 with or without common general knowledge, since none of the documents pursues such an objective and therefore there are no indications that the features of claim 1, namely ..., solve this problem.

Therefore, the subject matter of the new claim 1 is based on an inventive step, and consequently so are the subject matters of dependent claims 2 to 10.

Requests

It is therefore requested that a patent be granted on the basis of the new claims 1 to 10.

However, if the new patent application is not considered allowable and a rejection of the application is contemplated, an **oral hearing** is requested as an auxiliary measure.

The revision and adaptation of the description will be carried out as soon as allowable claims are available.

(Signature)

Annexes:

New set of claims

(Clean copy and marked-up copy)

MIX
Papier aus verantwortungsvollen Quellen
Paper from responsible sources
FSC® C105338

If you have any concerns about our products,
you can contact us on
ProductSafety@springernature.com

In case Publisher is established outside the EU,
the EU authorized representative is:
Springer Nature Customer Service Center GmbH
Europaplatz 3, 69115 Heidelberg, Germany

Printed by Libri Plureos GmbH
in Hamburg, Germany